Can Neuroscience Change Our Minds?

New Human Frontiers series

Harry Collins, *Are We All Scientific Experts Now?*

Everett Carl Dolman, *Can Science End War?*

Mike Hulme, *Can Science Fix Climate Change?*

Margaret Lock & Gisli Palsson, *Can Science Resolve the Nature/Nurture Debate?*

Hugh Pennington, *Have Bacteria Won?*

Hilary Rose & Steven Rose, *Can Neuroscience Change Our Minds?*

Can Neuroscience Change Our Minds?

HILARY ROSE AND
STEVEN ROSE

polity

First published in 2016 by Polity Press

Polity Press
65 Bridge Street
Cambridge CB2 1UR, UK

Polity Press
350 Main Street
Malden, MA 02148, USA

ISBN-13: 978-0-7456-8931-9
ISBN-13: 978-0-7456-8932-6 (pb)

A catalogue record for this book is available from the British Library.

Library of Congress Cataloging-in-Publication Data

Names: Rose, Hilary, 1935- author. | Rose, Steven P. R. (Steven Peter Russell), 1938-author.
Title: Can neuroscience change our minds? / Hilary Rose, Steven Rose.
Description: Cambridge, UK ; Malden, MA : Polity, 2016. | Includes bibliographical references and index.
Identifiers: LCCN 2015046128 (print) | LCCN 2016003260 (ebook) | ISBN 9780745689319 (hbk) | ISBN 9780745689326 (pbk) | ISBN 9780745689340 (Mobi) | ISBN 9780745689357 (Epub)
Subjects: LCSH: Neurosciences. | Brain.
Classification: LCC QP356 .R57 2016 (print) | LCC QP356 (ebook) | DDC 612.8–dc23
LC record available at http://lccn.loc.gov/2015046128

Typeset in 11 on 15 pt AGaramond
by Toppan Best-set Premedia Limited
Printed and bound in Great Britain by Clays Ltd, St. Ives PLC

For further information on Polity, visit our website: politybooks.com

CONTENTS

euroaesthetics, Neuroanatomy, Neuroanthroplogy, N
biology, Neurobiopolitics, Neurobollocks,Ne
ands,Neurochemistry, Neurocommunicat
eurocomputing, Neurocriminology, Neurocritic
euroculture, Neurodevelopment, Neurodiversity, N
economics, Neuroeducation, Neuroengineering, N
epistemology, Neuroessentialism, Neuroeth
eurogastronomy, Neuroinformatics, Neurohomeopa
euroimmunology, Neurolaw, Neurolinguistic progr
ing, Neuromanagement, Neuromania, Neuromarket
euromolecular gaze, Neuromorality, Neuromyth, N
pathology, Neurophenomenology, Neurophiloso
europlasticity, Neurophysiology, Neuropolitics, N
psychoanalysis, Neuropsychiatry, Neuropsychol
eurorobotics, Neuroschooling, Neurosexism, Neur
logy, Neurosoda, Neurotalk, Neurotheology, Neuro
l, Neurovirology, Neurowar, Neuroaesthe
euroanatomy, Neuroanthroplogy, Neurobiology, Ne
politics, Neurobollocks, Neurobrands, Neurochemi
eurocommunication, Neurocomputing, Neurocrimi
y, Neurocriticism, Neuroculture, Neurodevelopm
eurodiversity, Neuroeconomics, Neuroeducation, N
engineering, Neuroepistemology, Neuroessentiali
euroethics, Neurogastronomy, Neuroinformatics, N
homeopathy, Neuroimmunology, Neurolaw, Neuro
istic programming, Neuromanagement, Neuroma
euromarketing, Neuromolecular gaze, Neuromora

ACKNOWLEDGEMENTS

We wish to acknowledge the generous help our readers, Simon Gibbs, Maureen McNeil, Helen Roberts and Vince Walsh, have given us. That they come from the very different but very relevant disciplines of educational psychology, women's and cultural studies, medical sociology and human neuroscience has critically nourished our intellectual and political project of bringing the claims of neuroeducation into public scrutiny and debate. We have enjoyed debating the neurosciences and the theoretical frames through which they can be viewed with Steven's younger brother, the sociologist Nikolas Rose – especially over good food and wine. At Polity, Jonathan Skerrett, our perceptive editor, has been a pleasure to work with, as have indeed his colleagues. Apologies to those whose work we have drawn on but in this short book have had no space to acknowledge.

INTRODUCTION

The proliferating prefix

How to understand the ever-proliferating neuro-prefix, attached to everything from new academic disciplines – neuroeconomics, neuromarketing, neuroethics, neuroaesthetics, neuropsychoanalysis – to the marketing of neuro-eticals such as the soda drinks NeuroBliss, NeuroPassion etc.? Neuro occupies more and more space within mainstream science; neuro research papers dominate the leading journals, *Science* and *Nature*. Specialist journals proliferate. Neuro books, from the academic to the popular, stream from the presses. One publishing house alone, the venerable Oxford University Press, has no less than 1,200 neuro titles in its list. They range from handbooks on the wiring patterns of the brain through philosophical reflections on the relations between brain, mind and consciousness, to self-help texts on brain optimization. Unsurprisingly, the press has not been left out of this neuro-feeding frenzy; one research study mapped the steady rise of newspaper articles on the brain in the UK's top three broadsheets and the top three tabloids between 2000 and 2010.[1] The flow only slackened in 2008, when that year's near-cataclysmic banking crash briefly drove neuro from the pages. Brain optimization and brain pathologies, ranging from eating disorders to dementia, drew most of the journalists' interest. Neuro has gone into orbit and neuromania is all too often the order of the day.

Can neuroscience really change our minds? As a neuroscientist and a sociologist who share the view that neuroscience is dramatically increasing our understanding of the brain and also that science and society shape each other – that is, they are co-produced[2] – we have written this book together to unpick the hope from the hype of these neuro-prefixes, arising, as they do, as part of today's neoliberal political economy. The hopes that the neurosciences offer equal, even surpass, those of genomics at the time of the launch of the Human Genome Project in 1990, but with one crucial difference. Then one of the world's leading molecular biologists, James Watson, claimed that 'our fate lay in our genes', thus only geneticists could offer hope through their molecular biology, their genetic manipulation and their bespoke drugs, and could save us from our destiny.[3] The imaginary that they depicted left those outside the community of molecular biology as passive, waiting rescue. Despite the ideology of reductionism they share, the neurosciences' imaginary is radically different, claiming that their knowledge can empower us to remake our brains, and hence our minds and our very selves. Personal effort, guided by the neurosciences, can overcome the injuries of poverty and inequality. Plasticity, a property of the brain central to neuroscientific thinking for half a century, has become a quasi-magical term within public-policy discourse, offering an entirely new solution to problems of child development and poor educational performance, and heralded as the new elixir by the self-help manuals.

So is the answer to the question in the title of this book a simple 'yes'? As we will show in the chapters that follow, things aren't quite that simple.

For neuroscientists, the brain is the last biological frontier. It is seen as the repository of learning, thinking, deciding, acting, feeling angry, afraid, loving, remembering, forgetting, even consciousness[4] itself. Well funded, with €6 billion for just two Euro-American megaprojects, bolstered by an astonishing array of new technologies from the atomic to the systemic, and with research papers tumbling out in a seemingly inexhaustible torrent, it was almost inevitable that for most neuroscientists all doubts would vanish: the mind is the brain, the brain the mind. With this, the philosophical debate of centuries is simply bypassed.

Not everyone goes along with this, although an increasing corporatism in universities is hostile to dissent, and controversial ideas that might stir thought are unwelcome. With research money in short supply, only a handful of neuroscientists are willing to stick their heads above the parapet. More audible are the grumbles from psychiatry, although individuals who have vigorously entered public debate have encountered problems. British psychiatrist David Healy found his appointment to a senior position in a Canadian university blocked, following pressure from a pharmaceutical company, the efficacy of whose drug he had questioned.[5] Philosophers are freer from the constraints of needing substantial research grants, and many, John Searle, Raymond Tallis and Mary Midgley among them, have mounted a vigorous public defence of the mind.

The co-production of neuroscience, society and the self

Scholars working in science and technology studies – for the most part sociologists, anthropologists, philosophers and historians – have observed the fusion of science and technology in genomics, informatics and the neurosciences, and renamed them technosciences. Genomics would be impossible without high-throughput sequencers, and neuroscience without its imagers, and neither without mega-powerful computers.[6] The technosciences and today's neoliberal political economy are not separate entities: they are co-produced; the demands of the political economy shape the development of the technosciences, while in turn genomics and the neurosciences are powerful sources of innovation and hence provide the economic growth on which capitalism depends. What this structural account leaves out, however, is the agency of humans, both the technoscientists studying and manipulating life itself – from plants to animals, including the human animal – and the audiences and users of these new knowledges and technologies. The neuroscientists offer compelling imaginaries of how this new knowledge will help midwife new and hitherto undreamed of societies, while more mundanely their new understanding offers possibilities of manipulating the brain – from therapeutic interventions to new military neurotechnologies.

As so often with new technologies, humans modify their use for purposes other than were intended; as, for example, with the telephone, intended to make business more efficient but hijacked to facilitate social communication, call the family or speak to friends. The neurosciences offer similar possibilities. Anthropologist Rayna Rapp's ethnographic study[7] describes the experience of children and young adults going through hi-tech procedures to diagnose neurological problems, such as those of dyslexia or Asperger's syndrome. She observed that some, particularly young adults with a common diagnosis, came together as biosocial groups, not in opposition to neuroscience but seizing it as a resource to support their claim to a different, not deficit, brain identity. Their emphasis on diversity not deficit has found support from among the neuroscientific and clinical communities, and together they are building a new concept of neurodiversity. No longer are normal and abnormal brains set against each other; instead neurodiversity locates the neurotypical as one (albeit the most numerous) among many different brains.

This biosocial concept of neurodiversity offers more open ways to think about identity. By contrast, others, such as the philosophers Fernando Ortega and Francisco Vidal, argue that we live in a neuroculture, and thus conceptualize the brain as the centre of the self, as 'brainhood'.[8] Such theories of individual self-making exclude biosociality and hence the possibility of a shared construction of neuro-identity – a philosophical move which tends to reinforce the reductionist ideology of neuroscience, so clearly expressed in the titles of Jean-Pierre Changeux's book *Neuronal Man* or Joseph LeDoux's *Synaptic Self*. Such titles ignore Rapp's reminder that 'A child surrounds this brain.' Can brainhood make space for an identity of resistance?

The technosciences in neoliberalism

Since the mid-1970s the social rights embedded in the welfare states of Western Europe have been steadily eroded, a process dramatically accelerated by the near meltdown of the banking system. (The US trajectory has been rather different, never having had a welfare state on the European model – think of Obama's ongoing battle to secure access to health care for the poor.) Europe now follows the US and welfare is increasingly targeted towards the poorest by means-testing, even though research has long demonstrated that it is expensive to administer, humiliating to the recipients, and often misses those in greatest need. In Britain, even the right of free access to the treasured NHS is under systematic attack, beginning with the denial of health care to refugees.

In economics, it's been goodbye Keynesianism and hello to Chicagoan, or neo-classical, economics, with its reliance on complex algorithms and huge computing power. This latter, beginning as a distinctly marginal approach, rapidly increased its influence as cracks began to appear in the long post-1945 boom, with confidence in the welfare state fast fading. Today, despite the 2008 crash, which triggered a brief turn to Keynesianism, Chicagoan economics rises like the phoenix from what has unquestionably been ashes for most. The market remains fetishized as the guarantor of efficiency, innovation, economic growth and wealth creation, despite the challenge of the Occupy movement with its attack on the banking system and the obscene wealth of the 1 per cent.

Within this intensely marketized economy, social cohesion is weakened and collectivity fast replaced by the culture of what the political scientist C. B. Macpherson termed 'possessive individualism'. What then might be expected of the sciences co-produced with neoliberalism? When, in 1975, biologist E. O. Wilson published his foundational text, *Sociobiology*, its message was in accord with Macpherson's thesis. *Sociobiology* offers to explain why we – that is, humans – are what we are and do what we do. It draws on animal studies, genetics and evolutionary theory to argue that societies are indeed aggregates of selfish individuals, whose telos is the propagation of their genes into succeeding generations. By the 1990s, sociobiology, rebranded as evolutionary psychology, was offering a fully fledged account of human nature as universal, fixed in the distant past of the Pleistocene, and persisting ever since, across all societies and despite 200,000 years of social, cultural and technological change. Its conception of this genetically driven universal human nature is hierarchical, individualistic, competitive and patriarchal. In the world as conceived by evolutionary psychology, collectivity within a group – be it nation or state – is possible only insofar as it is genetically advantageous to the individual. Evolutionary psychology thus ideologically positioned itself against the welfare state, with its ideology of cooperation and universal social care.[9] As Wilson put it, humans might possibly create a fairer, more equal society, but only at the cost of losing efficiency.

The co-production of evolutionary psychology's theoretical apparatus and the ideology of neoliberalism is all too evident. But while it commanded substantial media space, it took little more than a pittance from the life-science budgets. The big money was increasingly in the biomedical technosciences, above all genomics and the neurosciences, seen as not merely wealth creators through innovation, but also as elegantly tailored to neoliberalism's shift from the collective to the individual. Neuroscience's preoccupation with the workings of the individual brain, even when the owner of that brain is engaged in intensely social interaction, and its reduction of persons to collections of neurons (nerve cells) and synapses (the junctions between them) is thus in accord with this focus on the individual, each 'neuro-self' responsible for their own well-being, sustained through the promises of personalized medical care.

In the chapters that follow, focusing mainly on the UK, we examine the ways in which, within this neoliberalism, neuroscience is being called upon to shape social and educational policy, targeting the deprived and the unemployed, who are blamed for what is described as poor parenting and thus limiting the 'mental capital' and aspirations of their children (Chapter 3), while offering the prospect of rational neuroscience-based education to enhance and optimize the brains of the young (Chapter 4). But, first, we ask how, from the dreams of an infant science half a century ago, neuroscience rose to its current pre-eminence (Chapters 1 and 2).

ONE

The Rise and Rise of the Neurosciences

The genealogy of the neurosciences

Of course, the development of the brain sciences long pre-dates the current neuro-surge, although today's flowering matches, if only coincidentally, the late twentieth- and early twenty-first-century rise of neoliberalism. But it is worth sketching out, if only briefly, this early history, beginning as it does for Western science with Descartes' location of the pineal gland in the brain as the junction point between soul and body in the 1630s – the moment when the brain became the site at which two different traditions converged: philosophers interested in the workings of the mind and the seat of the soul; and biomedicine, interested in the functions of the brain in the bioeconomy of the body and its various pathologies.

It took another two centuries for a new science to offer a materialist account relating brain to mind:

phrenology, seen then as a science but now derided as quackery, claimed to be able to infer people's temperament, proclivities and abilities through feeling the bumps on the surface of their skull. (Today's brain imaging with its renewed search for location has often been teasingly referenced as internal phrenology.) Phrenology apart, all that was available to the early researchers trying to relate structure to function were dead human brains. How much could be achieved within this limitation was demonstrated by the French anatomist Paul Broca, in 1861. By dissecting the brain of a recently dead patient who had lost the power of speech, he located a lesion in the left hemisphere, enabling him to conclude that this region was a speech centre. Broca's achievement fostered an enthusiasm for dissecting the brains of famous or notorious individuals, with a view to seeking the brain locus of genius or criminality.

As surrogates for brains, anatomists and physical anthropologists collected skulls, first privately, and then consolidated into vast collections in the natural history museums of Europe's major cities. Stephen Jay Gould has meticulously documented how, as European men of their time, the collectors knew even before they started measuring that the skulls in their collections would bear the stamp of the 'natural' hierarchy. A mass of critical historical research has analysed how, at the macro-level of the co-production of science and society, biology's normal brain was shaped by nineteenth-century imperialism and patriarchal social relations. Hence the brain of the rising middle-class white male was constructed as the standard of normality. The rest – organized by combinations of gender, class and colour – were arranged in hierarchical order, with the necessary subordination of women and people of colour, above all the black woman, to the white male. Eminent biologists such as Harvard's Louis Agassiz argued that women's self-evident mental inferiority required their exclusion from university study for their own protection. The stress of academic study on their weak brains would inexorably weaken, even destroy, their primary biological and social role of motherhood; their other role of the sexual servicing of their husbands was left in discreet silence.

So convinced were the researchers that brains could yield the secrets of mind that, in the 1920s, the young Soviet Union built an entire institute in Moscow for the study of Lenin's brain, sliced into sections for microscope analysis. (The Moscow Institute also houses a dolphin brain, besides whose complex convolutions even Lenin's retreats.) By comparison, the study of Einstein's brain was a distinctly amateurish affair; removed by his doctor at his death in 1955, it was cut into small blocks and stored in mayonnaise jars, before being distributed to interested colleagues. While little of value came from such microscopic observations of the brains of geniuses, the pathological approach which correlated brain damage with loss of specific functions began to establish a map of the brain, locating speech, vision, motor control and other capacities to specific brain regions. As with so many other areas of science, war (especially those of the late nineteenth and early twentieth centuries), with its heavy toll of brain-damaged young men, greatly speeded up such mapping.

Dead and damaged brains were fine for anatomists and microscopists, but, to study the workings of the living brain, physiologists and biochemists turned to the well-established laboratory stalwarts – rats, cats, dogs and occasionally monkeys – with all the problems of translation between animal and human brains that this entails. In animals, physiologists could study the electrical properties of nerves, biochemists the specificities of brain chemistry and metabolism, and pharmacologists the effects of drugs on both. These reductionist methodologies allowed some to follow in the footsteps of their most energetic nineteenth-century materialist precursors in dismissing the mind as a mere manifestation of underlying brain processes. However, most brain researchers were more modest, agreeing with the early twentieth-century physiologist Charles Sherrington in emphasizing the limits to their science, which could follow the workings of the nervous system through the deep structures of the brain only so far, but could not penetrate the mysteries of the cortex, whose grey matter was assumed to harbour the mind. Mind, consciousness and human behaviour were considered the province of philosophers or psychologists, who, especially under the influence of Pavlov and Skinner, tended to treat the brain as a black box, a machine for processing sensory inputs and outputting behaviours, but whose internal mechanics were a matter of indifference – much as most of us drive our cars.

The birth of a new science

Over the first half of the twentieth century, the several neurodisciplines established themselves in the academic hierarchy, each with its own problematic, methods and standards of proof, its own career structure, professional society and university department jealously guarding its intellectual and institutional territory. To transcend these disciplinary boundaries, to trespass on the territory of mind, was to risk being dismissed as unsound, a philosopher rather than a scientist. But, by the 1960s, with an expanding economy and an increasing science budget, a handful of visionary researchers decided that the time was ripe for a new research programme that would bring together all the disparate disciplines directed towards the study of brain and behaviour. The visionaries, a group at MIT led by the physiologist Frank Schmitt, did just this, with their Neuroscience Research Program. Neuroscience was exciting, more of its time than the archaic-sounding brain research. Neuroscience captured both the scientific imagination and the backing of the big US funders, both essential for the success of any research programme. It was to be the science of the brain, irrespective of the level, from the molecular to the systems, at which it was studied, and it was to be catholic in the choice of technique. What mattered was the project, to unite the disciplines in search of a coherent theory of how the brain works – and indeed, to override Sherringtonian hesitations by discovering how its neurons and their connections generated sensation, memory, the self, or mind. To use today's language, neuroscience is a big tent, and Schmitt and his colleagues were happy to open its doors to brain modellers, mathematicians, psychiatrists and psychologists, though it took another three decades for psychoanalysts to clamour to enter, seeing that neuroscience offered the prospect of anchoring their theoretical claims within the more prestigious life sciences.

Schmitt's project quickly drew in young researchers. Less caught up in the institutional structures of the existing disciplines, they were intellectually excited by its research programme and encouraged by the substantial funds it was securing. Within its embrace, new prefixes appeared; anatomy became neuroanatomy, biochemistry neurobiochemistry, physiology neurophysiology, psychology neuropsychology. Always receptive to the most challenging areas of science, leading scientists from other fields began to move into the brain. The decades following the discovery of the double-helical structure of DNA in 1953 had transformed genetics, and for many of the most theoretically driven molecular biologists it seemed as if all that was left to do was a little tidying up of loose ends. They were content to leave sequencing the genome, building DNA biobanks and developing the biotechnology industry to others. For them, as for the young enthusiasts, the biggest and unresolved challenge in biology was the brain. So it was into brain research that molecular biology's superstars like Nobel prizewinners Francis Crick and Gerald Edelman moved, in the firm conviction that the reductionist methods and molecular insights that had helped them solve genetics and immunology were the right ones with which to solve the brain, and with it the greatest prize of all – consciousness.

All the elements needed for a new research programme – visionaries, major theoreticians and young innovative researchers, and the mundane but essential funding – had thus been assembled. Neuroscience never looked back. From its foundation with a few hundred members in 1969, the US Society for Neuroscience has grown to a mighty behemoth whose annual meetings, though dominated by the US, attract some 40,000 researchers from across the globe. By the 1990s, the claims of the neuroscientists, their science transformed by the power of the new molecular and digital technologies into a fully fledged neurotechnoscience, had become so ambitious in their imaginaries and their attendant funding that both the US and Europe announced that this was to be the Decade of the Brain. The researchers themselves were confident enough of their own success to predict that at the millennium the brain decade would seamlessly segue into the Decade of the Mind. Reductionism would triumph; as Crick provocatively put it, locating the seat of free will in the anterior cingulate sulcus, a frontal region of the brain's cortex activated when a person is trying to solve complex problems, 'You are nothing but a bunch of neurons.'[2]

There's a macho toughness about Crick's language, celebrating what one of the founding members of the Molecular and Cellular Cognition Society (*sic*) described as 'ruthless reductionism'. Such fundamentalist reductionism is common among molecular neuroscientists (rather less so among neurophysiologists and psychologists), though few would put it quite so bluntly. This claim, powered up by the huge apparatus of today's technosciences, interrupts that of philosophy, where for two millennia philosophers have seen their deliberations on the mind as occupying intellectual centre stage. For the neuroscientists of the twenty-first century, mental activity can be reduced to brain processes, the continually fluctuating flow of neurotransmitters between the hundred trillion connections that connect the neurons of the human brain. The task of their science is thus to elucidate the genetics, biochemistry and physiology of these brain processes and, in doing so, to make the mind, and the person it inhabits, merely 'a user illusion', fooling people into thinking that they are making decisions whereas it is really the brain that is doing it.

Some philosophers, mainly in the US, have gone along with this, rebranding themselves with one of the proliferating 'neuro' prefixes, as neurophilosophers, for whom any talk of reasons and intentions – even consciousness itself – is no more than 'folk psychology' to be replaced by the austere formulae of computational neuroscience. Love, anger, pain, moral feelings – all are nothing but software within *The Computational Brain* (Churchland). The titles of neuroscientists' popular books express this increasingly molecularized and digitalized gaze, from *The Ethical Brain* (Gazzaniga) to *The Tell-tale Brain* (Ramachandran), *The Emotional Brain* (LeDoux) and *The Sexual Brain* (LeVay). Such titles are certainly constructed to catch the potential reader's, and better still customer's, eye but they also speak of the taken-for-granted present-day zeitgeist.

The power of neuro

Nothing could have been further from the intentions of the pioneering elite amongst the neuroscientists of the 1960s than dynamizing the economy or fostering new therapies for the diseases of brain and mind – at least, according to their own accounts. For them, their research was curiosity-driven. But the term 'curiosity' ignores the insistence of Francis Bacon at the birth of modern science that knowledge – for him, scientific knowledge – is power: power to control both inanimate and animate nature. The discovery of the structure of

DNA in 1953 is a good example. The finding was recognized as revolutionary by molecular biologists and geneticists, but passed without comment in the general media, even though within thirty years this newly identified structure came to frame the understanding of the living world and genetically manipulate it.

For the new generation of neuroscientists, it was to be the knowledge of the brain and hence the mind that would provide the power. For those with immediately practical agendas, notably a fast-expanding pharmaceutical industry seeking new psychotropic drugs, and the ever-watchful US Defense Advanced Research Projects Agency, DARPA, there was much to hope for from a field that was advancing so rapidly. Chemistry and physics had been militarized since the First World War, and with increasing emphasis since the Manhattan Project that had produced the atomic bomb in 1945. By the 1960s and the Vietnam War, it was biology's turn to enter weapon production, and neuro had its part to play.

For the military researchers at the height of the cold war, the new drugs that were spilling out from the labs offered chemical weapons to attack the central nervous system. Enthusiastic neuroscientists and psychologists were recruited into exploring the potential of covertly administered LSD to suspected spies, and even the

CIA's own agents, in one case causing one to jump out of the window. The US Chemical Corps stockpiled a nerve agent codenamed BZ, which, sprayed on to enemy forces, was supposed to disorient them and make them throw down their weapons in fits of helpless laughter (if only!). Advances in informatics suggested that brain processes could be modelled, or even enhanced, in computers, and, along with the precursor to the internet, ARPA (the D came later) funded a massive programme in artificial intelligence that laid the foundation for the present-day mega-brain projects. (DARPA's AI programme was, from its inception in the 1950s, plagued with the same theoretical fights between top-down and bottom-up modellers that have dogged the European Human Brain project, to which we turn in the next chapter.)

Hunting the molecule of madness

With the growth of neuroscience came a move to anchor psychiatry in neurochemistry, built on the underlying assumption – rooted in much older conceptualizations of madness – that the unbalanced mind was caused by a chemical imbalance in the brain. The task was to discover, as one leading researcher put it, 'how a disordered molecule leads to a diseased mind', and for pharmacologists (now *psycho*pharmacologists) to identify the chemicals that would restore both brain and mind, a programme vigorously fostered by the pharmaceutical industry.

The goal of matching particular molecules to particular psychiatric diagnoses underpinned the classifications of the *Diagnostic and Statistical Manual* (*DSM*) of the American Psychiatric Association, first published in the 1950s. One consequence was a rapid increase in the numbers of those deemed, according to the *DSM* criteria, to be psychiatric casualties – depressed, anxious, manic, schizophrenic – thus incidentally increasing the potential patient market. This rise in the medicalization of everyday distress consequent on job loss, bereavement, divorce and the many other adversities of daily life has continued unabated through the subsequent quarter-century, a process facilitated by the close relationship between the American Psychiatric Association, health insurance and the pharmaceutical industry.[3]

The underlying theoretical and practical problem, which remains to the present day, though cheerfully ignored by psychopharmacologists and many biological psychiatrists, is how to relate the classifications of the *DSM* – which are essentially phenomenological, based on listening to and observing the patient – to the assumed neurochemical causes. There were – and still are – no neurochemical markers to match against the *DSM* diagnoses.

The argument that we construct our self-identity from neuroscience is specific to the dominant neuroscience of the day. Thus, when neurochemistry was the dominant paradigm, it made sense to think of a 'neurochemical self'. As neuroscience's paradigms changed with the advent of powerful new technologies, above all imaging, neurochemistry has become but one of many neuroscience resources from which personal identity may be constructed.

With no neurochemical markers, many of the first generation of modern psychotropic drugs were discovered accidentally, having been originally synthesized for some other condition. For example, the world's first anti-psychotic drug, chlorpromazine, better known as Largactil (Thorazine in the US), was originally produced in the 1950s by Rhone-Poulenc Chemicals as an aid to anaesthesia; its sedating effects were only later recruited for the care of psychiatric patients. It became the first of the psychotropic blockbusters, prescribed for schizophrenia to some 50 million people over its first decade despite accumulating evidence of its severe adverse consequences – a movement disorder known medically as tardive dyskinesia and colloquially as the

Largactil shuffle. By the 1960s, chlorpromazine had been joined by the first of the tricyclic antidepressants, imipramine, and Hoffman-LaRoche's anxiety-reducing diazepam (Valium, Librium), the top-selling pharmaceutical in the US between 1969 and 1982, with sales peaking at 2.3 billion tablets in 1978. As 'mothers' little helpers', and prescribed as the ideal drug for women at the menopause, they were vigorously marketed to both doctors and, in the US, to the general public. Over-prescription led to widespread problems of dependency and addiction.

How and why the psychotropics exerted their effects on brain chemistry remained uncertain, though a consensus rapidly developed, on the basis of animal studies, that they worked by interacting with neurotransmitters – the chemical messengers that carry signals from one neuron to another across the synapses, that is, the junctions between the neurons. Chlorpromazine itself interacts with several such transmitters, but primarily dopamine. The finding led to the assumption that psychiatric disorders in general, from depression and anxiety to schizophrenia, were caused by disorders of one or more of the transmitters. Back in the 1950s, neurochemists assumed that there were only three or four major transmitters, each with a single mode of action. It is now clear that this was a huge oversimplification; there are multiple subtypes for each of the transmitters, and each in turn interacts with different receptor molecules in the synaptic membrane, and with the different enzymes that synthesize or break down the transmitter molecule itself. There are thus many ways in which a potential drug may interact with, boost or block the workings of the transmitter molecule.

In the three decades following the 1960s, as more and more transmitters and their related enzymes were discovered, each became in turn the fashionable molecule of the moment, a putative unique source of the neuromolecular discontent believed to underlie psychic distress. In the absence of better theory, the drug industry resorted to what was widely spoken of as 'molecular roulette', synthesizing thousands of molecular variants in the hope that they would discover the precise magic bullet with which to treat the distress. This crassly empiricist approach resulted in the patenting of a cascade of new drugs, each with massive hype, affecting not just dopamine, but other major neurotransmitters, acetylcholine, gamma-amino-butyric acid, serotonin, their subtypes, receptors and associated enzymes. It was one of the latter, Eli Lilly's specific serotonin reuptake inhibitor (SSRI) Prozac, that was celebrated in the 1990s as the happiness drug, making those who took it 'better than well . . . more like themselves'. Prozac became the world's most popular antidepressant, with 650,000 prescriptions being written each month; by 1990, it was generating $350 million a year for Lilly even as evidence mounted of the adverse reactions – including violence and suicide – to the drug and its rival, GlaxoSmithKline's Paxil. Sales figures for the psychotropics in general ran as high as $76 billion annually in the 1990s, until their patents expired and prices dropped. Throughout these boom years, the giants of the drug companies – Big Pharma – have been a potent source of funding for molecular neuroscience.

The animal model and its limits

The hypothesis that psychic distress results from a transmitter disorder and that the psychotropics work by correcting that disorder opened the way to a more systematic, laboratory-based search for new drugs. Interfering with the workings of the transmitters in lab animals should, on this thesis, somehow mimic the schizophrenic, anxious or depressed behaviour of humans, and the animals could thus serve as test beds both for theories of human psychiatry and the development of new drugs. Such animal models became the standard approach to psychotropic drug discovery. By the millennium, advances in genetic technology made the animal models even more attractive. Mice can now be genetically manipulated ('constructed' is the preferred euphemism), specific genes deleted and others inserted, activated or deactivated almost at will. For example, human genes predisposing to Alzheimer's disease can be inserted into mice and the animals studied for signs of the changes in brain chemistry and memory loss seen as characteristic of the disease in humans.

The problems with such animal models, especially for the psychiatric as opposed to unequivocally neurological disorders, are manifest. What sort of behaviour in a cage-reared mouse might be seen as analogous to schizophrenia in a human? Is a rat or monkey sitting motionless in the artificial context of its cage, oblivious to food or sexual partners, a model for human depression? How can a mouse's loss of memory for the route through a maze be compared to the age-related loss of memory for names described, for instance, in Penelope Lively's autobiographical *Ammonites and Flying Fish*? And how can such animal models be matched against the epidemiology of depression (twice as frequently diagnosed in women as in men) or schizophrenia (twice as common in working class than middle class, and commoner in Britain in people of Caribbean than European ancestry). Only in 1993, under steady pressure from feminists, were women of childbearing age included in clinical trials of new drugs in the US, and effects controlled for what were seen as sex and race differences. For decades, the standard animal model was a male rat or mouse, on the grounds that female cycling confused the results. It wasn't until 2014 that, in the US, the major federal funder of academic research, the National Institutes of Health, decreed that henceforward animal studies must involve equal numbers of males and females. But by then both animal models and psychopharmacology were in some trouble – trouble compounded by a paper showing that lab rats respond differently to male and female scientists.[4]

The seeming success stories of the 1980s and 1990s have not been replicated in the decades that followed. Psychic distress is, it is claimed, increasing, with the World Health Organization speaking of an epidemic of depression sweeping the globe, reflecting the medicalization of what in other societies or only a couple of decades ago in Britain would be regarded as unhappiness. Thus, the latest, fifth edition of the *DSM*, published in 2013, regards grieving over a dead partner or child lasting longer than a fortnight as indicative of clinical depression. Also increasing in the ageing populations across the industrial world is the diagnosis of dementia, said to affect 800,000 people – predominantly women – in the UK. The number is projected to rise to a million by 2021. The causes of this rise are multiple and complex; the isolation, loneliness and poverty that is the lot of so many older people in Britain today reflect a growing crisis for public health and social care in a fragmented and individualizing society. A person's responses to such isolation may be diagnosed as dementia, even without a specific neurological cause such as Alzheimer's disease. The Alzheimer Society reckons dementia costs the UK £26 billion a year – a figure which ignores the unpaid work of carers. And despite the detailed knowledge of the genetics and biochemistry of the disease, and intense research efforts, there are still no new and effective drugs that can do more than briefly slow the development of the disease.

In 2011, in a bid for enhanced funding for neuroscience, the European Brain Council calculated that in Europe 38 per cent of the population – 165 million people – would develop mental illness in any one year at an annual cost of €800 billion or 24 per cent of the continent's health budget, though much of this disease, the Council conceded, goes unrecognized and untreated, so it is hard to know how much weight the figures can bear. Perhaps they can best be regarded as estimates of the potential size of the market for the psychotropics. This ought to be a bonus for the pharmaceutical industry. But the patents are running out, there are no new blockbuster psychotropic drugs in sight, and those that have been developed in recent decades – even the much-vaunted SSRIs – are widely conceded to work no better than their predecessors of the 1960s.

By 2010, major companies like Pfizer and GlaxoSmithKline began closing down their traditional neuroscience research labs, initially refocusing funding towards the seemingly more productive terrain of cancer and heart disease. The fundamental problem is that the straightforward causal chain connecting disordered transmitters with diseased minds does not hold. The animal models are, to say the least, inadequate. Many would question even the possibility of matching the phenomenological account of being depressed or anxious with a neat biological correlate – so much so that the latest, 2013, edition of the *DSM* ran into a barrage of criticism from psychiatrists and neuroscientists alike. From now on, decreed Steven Hyman, 2014 president of the American Society for Neuroscience and past director of the National Institute of Mental Health, it should only fund research directed towards a clear biological – ideally genetic – target, a biomarker which could reliably predict treatment response.[5] Perhaps not coincidentally, Pfizer, Novartis and others of Big Pharma are relocating their research labs in closer proximity to Hyman's own Harvard base. Meanwhile, psychiatrists will continue to reach for their prescription pads.

The happy marriage?

With biological psychiatry in clinical trouble, it has been the new technologies, notably fMRI and a family of related imaging techniques, which have transformed and reinvigorated the neurosciences by freeing them from the limitations of animal models through entirely new and previously unimaginable, non-invasive ways of studying the workings of the living human brain. It's not only the whole brain that can be imaged, but even individual nerve cells, which can be genetically marked to indicate when they are active.

With fMRI and related imaging techniques, it became possible to integrate the lab study of human behaviour – previously the province of psychologists – with what goes on inside the head, the neuronal pathways and brain regions involved in thinking, feeling and doing. New research fields, such as cognitive and social

fMRI

The starting point was the invention of magnetic-resonance imaging (MRI), a scanning method that involves positioning a person in a strong oscillating magnetic field. The field excites hydrogen atoms in the body (mainly in the form of water) and these in turn emit a radio signal that the detector picks up. MRI provides three-dimensional X-ray-type pictures of brain structures, invaluable for identifying damaged regions after stroke or trauma. The key development transforming such static images in dynamic movies of the brain at work came in the 1990s with functional magnetic-resonance imaging (fMRI). The brain uses a prodigious proportion of the body's oxygen, carried in the blood. Oxygen too can be persuaded by strong magnetic fields to emit a radio frequency signal. The oxygen level in the blood is measured by fMRI as the blood flows through the brain, which is taken as a surrogate measure of brain activity. The higher the blood flow in any region, the more active that region is assumed to be. As the brain is always active, the experimental designs involve comparing the blood flow, and hence oxygen use, through a person's brain during some mental task, such as identifying the odd one out from a word list, with that when they are at rest. If the flow increases in any region during the task, then that region is assumed to be necessary for – or even the brain site of – that activity.

neuroscience, were born. So optimistic were the imagers that they concluded, as one put it, 'fMRI offers a happy marriage between mind and brain'. Perhaps even more importantly, the beautiful false colour images that fMRI generates are read, not only by the general public, but also by the researchers themselves, as unmediated indicators of brain function. Brightly glowing pictures showing the brain regions active when the brain's owner attempts to solve a maths problem – or when a London taxi driver is asked to plot a route across the capital's notoriously tricky streets – have proved irresistible to the popular media. In an internalized reprise of phrenology, fMRI was claimed to be able to demonstrate brain sites for everything, from romantic love to economic choice and moral values, as if these were located in nature, not historically produced.

No matter that the fMRI images are far from being veridical snapshots of what is actually going on inside any given brain, but are deduced by a chain of manipulations and statistical assumptions to produce the dramatic pictures that give them their instant appeal; they are understood as real, and not only by non-neuroscientists. Such images can hide as much as they reveal. To start with, their timescale, of blood flow averaged over seconds, is too long when brain processes operate at millisecond speeds. So too is their spatial resolution. A cubic millimetre may not sound big, but in the cortex – the brain's grey matter – it contains no less than 5 million of the brain's hundred billion neurons, connected to one another and the outside world by way of 50 billion of its hundred trillion synapses, 22 kilometres of dendrites and 220 kilometres of axons.[6] The doubts expressed by Sherrington nearly a century ago about science's ability to penetrate the workings of the almost inconceivable complexity of the cortex should still be with us.

Even so, it is hard to resist the appeal of the fMRI images; the enhanced colour, the sense of watching the brain actually thinking and feeling. They are accessible and easy to understand in a way that those earlier more abstract images of genetic certainty, the barcode of sequenced DNA, even the representations of the double helix, were not. Although the rhetoric of some virtue or behavioural trait being 'in our DNA' persists – even proliferates – by the second decade of this century the fMRI images of the brain have become more deeply personal: 'my fMRI is me'. While very few in today's Photoshopping world would go along with the old saying that the camera cannot lie, we still seem to believe that truth claim where internal image-taking is concerned. Even neuroscience students are more likely to accept a false claim about the brain if accompanied by an irrelevant fMRI image. So powerful is the technology that more disturbing uses of brain imaging are being developed. Suggestions that questioning a person under fMRI can detect whether they are telling the truth or lying have led to proposals that imaging offers a more reliable form of lie detector than the traditional polygraph, which works by measuring skin conductance, said to increase when a person lies, and which – although still used in the United States – is known to be wildly inaccurate. Could fMRI be used in criminal court proceedings? In interrogating prisoners or alleged terrorists or spies? US companies like NoLieMRI have sprung up to exploit the technique, offering their services to the law and military despite serious doubts about its accuracy.

By the second decade of this century, neuroscience seems finally to have come of age. It was time for it to go mega.

TWO

The Neurosciences Go Mega

Although the 1990s had been titled the Decade of the Brain, mega-science projects, long familiar in physics and astronomy, were slow to arrive in the neurosciences. The first giant life science project had been built on the revolution in genetics with the plan to sequence the 3 billion DNA bases of the human genome, at a neatly matching cost of $3 billion. To convince potential backers that the human genome project (HGP) could and would deliver, its advocates invoked the success of the gigantic 1940s Manhattan Project with its delivery of the atomic bomb. To win public and political support, the molecular biologists claimed that sequencing the genome would lead not only to new scientific knowledge, but also the power to generate new gene therapies and tailor-made drugs for hitherto intractable diseases, and so create both health and wealth – a much more complex and diffuse set of objectives than the delivery of the biggest and most devastating bomb in history.

Three billion dollars was an unprecedented sum for a life science project. Funded primarily by the US government and the UK's Wellcome Trust (then the world's richest non-governmental body funding biomedical research), the public effort was under heavy competition from a rival, commercially funded company hoping to profit by patenting the gene sequences. Under such pressure, the sequence was completed by the millennium. The race itself fused genetic science with digital technology, hugely speeding up genetic and sequencing methods. In one sense, therefore, the molecular biologists did deliver what they had promised – the fully sequenced human genome. And the public investment in the project galvanized the biotech industry with a range of new products, from ultra-rapid gene sequencers to companies offering to read customers' genetic risk, thus feeding the neoliberal economy's insatiable greed for growth.

However, the completion of the project did not deliver health benefits on anything like the promised scale, and it is only now, nearly two decades later, that new gene-based treatments for cancers are beginning to be developed. The unexpected complexities in the genome which the project revealed (and which we have written about elsewhere[1]) have, however, resulted in a major shift in biologists' understanding of gene function. Perhaps if the HGP's most energetic advocates had eased up on the hype that it would lead to personalized medicine and new genetic therapies which would cure everything from cancer and Alzheimer's to schizophrenia, but instead had explained that completing the genome would provide a powerful platform for a further stage of research and that it was only then that any deliverables could arrive, they would not have raised – and then dashed – such high hopes.

The Human Brain Project

The Decade of the Brain ran in parallel with the HGP throughout the 1990s, but with a less visible public image. Nonetheless state funding for the neurosciences substantially increased in both the US and Europe. Future forecasting projects on both sides of the Atlantic saw the neurosciences as a potential major growth area, ripe for wealth creation, through the development of new psychotropic drugs and neurotechnologies. However, it wasn't until 2013 that neuroscientists lobbied successfully for their own mega-projects – and then not just one but two, one European, one American. Once more, the Manhattan Project was invoked; brain research could and would deliver. First up were the Europeans. The EU's €1.2 billion Human Brain Project (HBP) to decode the brain and create a brain-like computer was one of the two winners of the 'grand challenge' competition awarded under its Flagship Future and Emerging Technologies Programme. (The other winner was the Graphene Initiative, a new form of carbon with potential applications in fields as diverse as biological engineering and photovoltaics.) Significantly, the funds for the HBP came not from the European Commission's research arm, but from its Directorate on Information and Computer Technology – a source which makes clear that, whatever the claims of its advocates, its funders see the HBP not so much as a neuroscience project but one generating new computing technologies: 'A major driver of ICT in Europe', as leading researcher Thomas Lippert of Europe's supercomputing centre put it.[2]

The banking crash of 2008 brutally exposed the weakness of both the European and the North American economies, above all the sluggish growth of the real economy central to capitalism's future. The EU responded by selecting the two projects, graphene and the brain, as the technosciences most likely to restore growth. As with the genome project before it, there was little or no debate about whether government and society could afford to focus solely on economic growth, ignoring the issues of climate change and environmental harm. It is as if the German sociologist Ulrich Beck had not published his influential book *The Risk Society* in 1986, highlighting dangers brought about by unfettered science and technology. Nor, for that matter, was there any discussion of whether such technological developments should contribute to social growth, or whether the EU's huge research projects might exacerbate the growing inequalities of neoliberalism.

Many economists are sceptical of the nation-state or the EU funding the initial risky stages of research, leaving private industrialists to enjoy the profit-making stage. In today's global economy, neither the nation-state nor the EU can be certain whether it is even their own capitalists who will profit. Graphene was discovered in the UK by two scientists, both migrants from Russia, jointly awarded the 2010 Nobel Prize. Yet despite the European Commission's €1 billion Graphene Initiative in 2013, two years later the graphene market is dominated by China, registering four times as many patents and manufacturing twelve times as much graphene as Europe.

Public engagement in science

The decision to fund the HBP emerged from the closed circles of the Commission without consultation either with the wider European neuroscience community or the public. The failure to consult the neuroscientists more widely led to the open revolt we discuss in the rest of this chapter. But it was the failure to consult the public that highlighted a major U-turn in policy. Since the rise of the green movement in the 1980s – challenging technoscience's preoccupation with technological progress, above all genetically modified organisms (GMOs), and hence the failure to take care of the biosphere – public consultation has been recognized as a necessity. The continuing public hostility to GMOs led in 2015 to half the member states opting out of the EU's relaxation of its ban on such crops.

In Britain, not only was there hostility to GMOs, but the public became increasingly distrustful following the outbreak of Mad Cow Disease, as well as the claim in a medical journal that the triple MMR vaccine was associated with autism. The British government, having presided over more science-related disasters than most, was advised by the Royal Society of the urgent need to restore trust. The first project assumed that if scientists talked directly to the public, explaining the facts of science, the public would understand and trust would be restored. To test the validity of this assumption, the Economic and Social Research Council funded a series of studies to explore how the public understood and trusted science when confronted by science-related risk.[3] These concluded that the old deference to science and scientists had weakened, and that with it went the scientific community's belief that they and their advice alone were sufficient to guide policy. Overcoming this required the engagement of scientists with the public to foster a science which was acceptable to both science and the public, through mutually respectful dialogue. This new approach of Public Engagement of Science was endorsed by a report of a House of Lords Select Committee on Science and Technology in 2001 and became the template for restoring citizens' trust in science.

Public Engagement was at its high tide across Europe in 2006 when *The European Citizens' Deliberation on Brain Science* presented its list of priorities for brain research to the Parliament. Not surprisingly, the EU publicity machine went into hyperdrive:

> For the first time, citizens of the European Union are in the driver's seat in debates that are shaping public

> policy. The field is brain science, an issue of such importance that it has inspired a unique two year [consultation] ... unprecedented opportunity to give ordinary people a role in guiding the EU ... in policy development in a complex scientific field ... a breakthrough in participatory government.[4]

The Commission had provided the funds for the consultation, channelling it through two foundations, but otherwise playing no active role.

The result was a list of thirty-seven citizens' recommendations to the EU Parliament and Commission. These ranged from increasing funding for 'basic and fundamental research on both healthy and sick brains' to the socially responsible cautions: 'avoiding medicalizing society ... [not] ... using brain research as a means of social control ... recognizing diversity and the needs of and respect for psychiatrically and neurologically damaged people' to increasing the transparency of research funding and establishing a pan-European ethical committee to scrutinize brain research.

The HBP's grand project of building a computer simulation of the human brain did not appear on the list. So much for being in the driver's seat and the breakthrough in participatory governance: instead of 'ordinary people' being involved, the HBP has outsourced the task of securing public trust by appointing professional bioethicists and/or sociologists to the project.

In the beginning was the mouse

The Human Brain Project begins with the premise that, as Swiss-based neuroscientist Henry Markram, its initiator and coordinator, puts it,[5] the human brain is 'the world's most sophisticated information processing machine', operating, however, on principles currently unknown but 'that seem to be completely different from those of conventional computers'. The project's goal, therefore, was 'to build an information computing technology infrastructure for neuroscience and brain related research in medicine and computing, catalysing a global collaborative effort to understand the human brain and ultimately to emulate its computational capabilities.'[6] That is, the intention is to invent new forms of more brain-like – so-called neuromorphic – computing, and to create a computer model of the entire human brain by 2023. The scale of the ambition of this decade-long collaboration that recruited some 113 separate research groups in twenty different countries, supported by the Commission with matching funding from the participating countries, thus dwarfs the pursuit of the 3 billion DNA bases of the HGP. But, as we shall see, the project was mired in controversy even before it was formally launched.

Markram's proposal was based on his prior collaboration with the giant US computer firm IBM, whose European base is in Lausanne. It was IBM's Deep Blue computer that had finally triumphed over chess master Gary Kasparov. Flushed with that success, in 2005 the company joined forces with Markram, supplying the Blue Gene supercomputer he needed for his Blue Brain project to build a 'realistic' model, incorporating all that was known of the human brain's anatomy, biochemistry and physiology, beginning with something less ambitious – a small section of the rodent brain. The HBP, initially coordinated from Lausanne but now moved to a new campus in Geneva, is essentially a hugely scaled-up version of Blue Brain, and its imaginary of the neuromorphic computer captured the European Commission's attention. But with IBM's feet so firmly under the table, the European IT industry will have to work hard to avoid being frozen out. Not many of the researchers recruited to the HBP were aware of the plans for a private Swiss-operated foundation charged with exploiting the commercial opportunities arising from the project.

The initial intention was to collate the vast body of existing data about the connections and chemistry of the brain, feed them into conventional computer systems and use them to build models of how the brain might work in, for instance, enabling vision or memory. 'The brain', in this account, is assumed to be the human brain, yet in practice much of the data is bound to come from laboratory animals. The biochemistry, physiology and fine anatomical structure of neurons in small mammals like rats and mice, even birds, has much in common with that in humans and is the basis for most of what is known about the cellular processes at work in the human brain. However, the failure, so common as to be almost the norm, of drugs that have been effective in animal trials to work in humans – Alzheimer's is a good example – urges caution concerning translation.

A few months after the EU's public launching of the HBP, US neuroscientists successfully won President Obama's support for a matching effort. Their initial proposal, not dissimilar to Markram's, was to map all the neural pathways and connections (called the connectome) between the 70 million neurons of the mouse brain, as surrogate for the human brain. The scale of the task and the grandiosity of the ambition is indicated by the fact that, in 2015, after six years of painstaking anatomical study, a team of US researchers was able to report that they had produced a complete map of a minuscule 1,500 cubic micrometres of the mouse brain – smaller than a grain of rice. And the mouse brain weighs around a three-thousandth of the human brain – though this didn't inhibit the press release announcing the publication from suggesting that it might reveal the origins of human mental disease.[7] Almost simultaneously, Markram announced to a blaze of publicity that the HBP had modelled a minute fraction of the connections in a region of the rat brain linked to the rat whiskers and, by stimulating these connections, they could make the computer-modelled whiskers twitch. Other neuroscientists remained sceptical.[8]

DARPA and the BRAIN Project

In response to the connectome call, Obama pledged an initial $3 billion to the BRAIN Project (Brain Research for Advancing Innovative Neurotechnologies) – a budget increased by 2014 to $4.5 billion. Despite the funding promise, Obama's proposal did not win immediate and universal praise from the neuroscience community, the commonest criticisms being that it was premature and over-grandiose. But whereas the EU's project emerged by fiat from the Commission, Obama's was open to negotiation with the many and varied goals of the neuroscientists, who have been able to widen BRAIN's objectives well beyond the connectome. The innovative technologies BRAIN funds, often seemingly remote from studies of the brain but of wide industrial potential, range from nanoparticles to optoelectronics. Unlike the HBP, the US project's acronym makes clear its technological and wealth-creating intentions. Significantly, BRAIN is being funded not only by the US federal agency and the National Institutes of Health, but also by the military through DARPA.

DARPA's interest is straightforwardly above all in the development of neuroprosthetics, computer-aided devices to treat the limping parade of brain- and mind-damaged young soldiers – 300,000 in the US since 2000 – returning from wars abroad. In these asymmetric wars many of the US and UK casualties result from the improvised explosive devices which blow up armoured vehicles. Even when protected by helmets those inside the vehicles suffer brain trauma, often developing long after the explosion itself, just as the brutality of the wars can leave lasting mental scars. Brain–computer interfaces and implants to the motor or visual regions of the cortex can potentially bypass paralysis caused by spinal injury, or blindness from damage to the retina or optic nerve. Magnetic or electrical brain stimulation might relieve post-traumatic stress disorder or compensate for memory loss. However, the military's goals are not simply therapeutic. DARPA is interested in the ways such prostheses might enhance perception for an intelligence analyst trying to interpret an aerial photograph, or to speed up a pilot's decision as to whether and when to launch a missile.

As is often the case, medical technologies developed for the military move into civilian practice too; in 2015, the NIH picked up on the prospect of the implants, fostering a series of BRAIN workshops to encourage commercial development and pressing the FDA to speed up its approval procedures for new therapeutics. But, as leading bioethicists have pointed out, this politically driven emphasis on haste has failed to confront the ethical problems.[9] The neurosciences face special challenges both scientifically and ethically. Overweening ambition sets aside the complexity and poor understanding of the brain, the weak evidence from clinical trials of deep brain stimulation for depression, and concerns that the implants may cause radical changes in behaviour that affect a patient's autonomy. All urge caution.

Solving the brain

Europe and America's Big Brain projects were launched with cascades of superlatives – they would be 'transformative', solving 'the mystery of the 3lb of matter that sits between our ears', an achievement beyond Manhattan, putting an astronaut on the Moon or finding the Higgs boson. With 'a map of the human brain we would come closer to developing treatments for everything from depression and post-traumatic stress to Alzheimer's and paralysis.'[10] Such claims thus unblushingly repeat those made at the launch of the HGP – which genomics has so far been unable to deliver. Like the genome project, the brain projects are not just about scientific understanding and health but also – indeed above all – supposed to be wealth generators. Announcing his project, Obama referred to the claim that each federal dollar invested in the genome project had yielded $141 to the US economy. BRAIN, he averred, would do no less. Not to be outdone, in 2014 Japan announced its own mega-brain project, and as we write, in spring 2015, China is said to be ready to announce an even larger one. Maybe such neuroprojects will join space programmes as the sites for national rivalry? And, indeed, not only national. The Allen Institute, funded

by Paul Allen, co-founder of Microsoft, has also announced its own mega-project, Big Neuron, to go alongside its earlier precursor brain projects, the connectome and the brain atlas.

On the face of it, then, the brain projects are indeed neuroscience's Manhattans. But look a little closer and, despite the claims of their protagonists, the analogy begins to wear thin. Making an atomic bomb, determining the 3 billion DNA sequence, even without the promised deliverables to health, are clear objectives. But what would it mean to 'solve the human brain'? What might such a solution look like? Ask the many neuroscientists recruited to the two projects and you are likely to get as many different answers – and diverse research proposals. Even before the European award was announced many neuroscientists – including some close to Markram in Lausanne – began to voice their scepticism about the possibility of modelling the brain 'in silico' when the fundamental principles on which it operates remain unknown.[11] The neurosciences simply have no equivalent to the theoretical physics underpinning the Manhattan Project, nor the molecular biology of the Human Genome Project.

A controversy which has divided researchers working on artificial intelligence since the field first emerged in the 1950s is whether it is really important to unravel the fine details of the biochemical processes underlying neural connectivity, or the functioning of synapses, in order to model the workings of the brain on which cognition depends? Or should these molecular mechanisms be taken for granted and the focus instead be on modelling the ways in which the multiple and richly connected sub-systems of the brain interact? That is, bottom-up or top-down modelling? DARPA, in its funding of artificial intelligence, opted for top-down. The fine internal structure of the brain didn't really matter, DARPA's modellers argued; instead one should focus on producing a model which, when fed with appropriate inputs, behaved in a 'brain-like' way in learning, remembering and outputting correct responses.

Markram's insistence on bottom-up modelling is not shared by many neuroscientists studying higher-level brain functions, above all cognition. For Stanislas Dehaene, a leading Paris-based cognitive scientist, it 'will fail to elucidate brain functions and diseases, much like a simulation of every feather on a bird would fail to clarify flight'.[12] The debate at core is about the appropriate level at which to research and explain the workings of the human brain.

To succeed, the HBP needs not only to ensure the quality of the data entered into the computer simulations, and the management of such vast quantities of data – already hard enough – but also to have testable hypotheses to explore. And this is exactly what is currently lacking. For many neuroscientists, the HBP isn't really about the brain at all, nor, despite its rhetoric, is it likely to solve the pressing problems of neurological or psychological disease. Whatever the initial excitement at the prospect of substantial research funding, the rumbles of discontent amongst participants finally broke into the open in 2014 when some 700 neuroscientists signed up to a letter to the EU Commission, criticizing both the science and the management of the project, above all what the critics called the 'radically premature' modelling of the brain from the molecular level upward, and its sidelining of cognitive science, one of neuroscience's best-developed fields. The letter pointed to the narrowing of its aims and methods and called for an urgent independent review of the project by external experts.

The initial reply, coming, significantly, not from the EU's Research Directorate but from Robert Madelin, Director-General for 'communications networks, content and technology', attempted to pacify the critics, agreeing that there was 'no single road map for understanding the human brain' and stressing that the HBP was collaborating with BRAIN.[13] Markram's response was less emollient, dismissing the criticisms as sour grapes, and reflecting that such doubts had dogged the early days of the genome project. In response, two senior European neuroscientists, Yves Fregnac and Gilles Laurent, both initially engaged with the project, went public on the attack, with a powerful article in *Nature* entitled 'Where is the brain in the Human Brain Project?'[14] The Commission had no choice but to set up an external mediation committee, whose report, in March 2015, largely vindicated the critics. The project fails, it concludes, not only in its governance but also in its scientific plan, particularly the core aim, the simulation of the entire brain, and in exaggerating its clinical potential. The strongest criticism was directed at the project's governance structure, which despite the trappings of accountability and a built-in ethics committee had left overall power in the hands of a three-person executive committee, and in particular Markram himself, who 'is not only a member of all decision-making, executive and management bodies within the HBP, but also chairs them … Furthermore he is a member of all the advisory boards and reports to them at the same time … he appoints the members of the management team, and leads the operational project

management.' As Fury the cat says to the mouse in *Alice in Wonderland*, 'I'll be judge I'll be jury, says cunning old Fury, I'll try the whole cause ...'

Despite this damning critique, Markram and his fellow directors are said to regret that it risks turning their visionary project into an average one; modelling the brain in a computer was the HBP's 'unique selling point'.[15] With this, the language of marketing works to replace the language of science.

A rip in neuroscience's big tent?

Despite the optimistic claims and the rich sources of funding, the neurosciences' problems run deeper than criticisms amongst the profession about particular projects. Underlying the doubts is uncertainty about what sort of science neuroscience really is – or indeed if it can ever be a unified field of study as opposed to a portmanteau term which conveniently bundles together many different and sometimes even contrasting scientific endeavours. Despite the power of the fMRI images to recruit believers, the core problem remains. There is no central 'theory of the brain'. As the clash between Dehaene and Markram over the direction of the Human Brain Project makes clear, there is no way to integrate

the various neurodisciplines, from the molecular to the systems, still less a way to project them outwards into the mysterious terrain of 'mind'. The 40,000 neuroscientists who meet annually at the Society for Neuroscience often seem to have no common language. Two books on our shelves, each about memory, one by a cognitive psychologist and the other by a molecular biologist, share few common references; even their understandings of what constitutes 'memory' differ. The neurosciences are data-rich and theory-poor.

The Human Brain Project's remit – to bring all this data into a common and computer-formatted store – falters when such a mountain of data has been retrieved from multiple sources, often irreproducible, with different and irreconcilable measurement techniques and experimental paradigms. GIGO – the IT experts' bugbear of Garbage In: Garbage Out – overshadows all such projects. Unsurprisingly, many of the 'deliverables' from neuroscience seem as remote as ever.

And yet neuro proliferates

Over the past three decades, with genetically constructed mice, fMRI and its fellow imaging techniques, neuroscience has at last come of age and joined the world of Big Science. Today, its cheerleaders incessantly invite us to see how we may enhance our own and our children's brains by better parenting and education, and how by lifestyle modification we can postpone or even avoid senescence. Unlike genetalk, mocked by historian Donna Haraway's 'Genes'R'Us',[16] with its strong determinism, in which averting fate comes from outside via genetic manipulation and new drugs, neurotalk emphasizes the idea that the brain is malleable, plastic in response to challenge. 'Plasticity', once a technical term used by neuroscientists to describe the capacity of the brain to modify itself in response to insult and experience, has become the trope of neurohope. Embraced by the self-help manuals and courses offered to teenagers and their stressed-out parents, plasticity is hope literally embedded in the brains of all, nothing less than 'with my plasticity I can remake myself'. The power and prospect of neuro is indeed the message that the funding agencies – in Europe, America and the rising scientific powers of China and its eastern neighbours – have taken on board. However, as we will argue in the chapters that follow, such 'evidence-based' neuroscientific advice assumes our brains are somehow separate from our embodied location in the complex intersections of the cultural, social, economic, historic and environmental.

So where does this leave the proliferating neuro-prefixes with which we opened this book? In general, it is those professions, fields and practices whose theoretical foundations are weak, at least according to the ideology of the natural sciences, which seek security in the neuro-prefix, now so attractive that it is attached to practices and theories that have moved far beyond the goals that drove Schmitt and his fellow pioneers to attempt to construct an integrative science of the workings of the brain and nervous system. Some of this neurotalk is little more than fashionable froth and transient opportunistic marketing. Others, like neuroaesthetics, seek, in the face of critical opposition, to break new theoretical ground by rooting art appreciation into brain processes and human evolutionary history. The concerns of this book, however, are the expanding ambitions of the mainstream neurosciences, shaping and being shaped by neoliberalism, and increasingly entering – even seeking to colonize – public policy in child development and education. And it is to these issues that the following chapters turn.

THREE

Early Intervention

Making the Most of Ourselves in the Twenty-first Century

In 2008, on the very cusp of the banking crash and the prolonged recession that followed, the Labour government published a Foresight Report entitled *Mental Capital and Wellbeing: Making the Most of Ourselves in the 21st Century.*[1] Foresight Reports were devised to help governments take the long view: to identify opportunities, risks and priorities over several decades ahead, based on a mix of scenario planning and consultations with a formidable number of experts. As indicated by Foresight's location – currently in the Office for Science and Technology, itself part of the Department for Business Innovation and Skills – its reports are primarily oriented towards wealth creation. Thus, despite the commitment of the title to 'making the most of ourselves' and embracing the entire life cycle, the report's headline assertion is unequivocal: 'Countries must learn how to capitalize on their citizens' cognitive resources if they are to prosper, both economically and socially. Early intervention will be the key.' And neuroscience is crucial to their project of changing for the better the minds of the young.

Three concepts are central in the report: mental capital, mental well-being and cognitive resource. The first includes cognitive ability, flexibility and efficiency at learning, social skills and resilience. Mental well-being is 'a dynamic state that refers to an individual's ability to develop their potential, work productively and creatively, build strong and positive relationships with others and contribute to their community.' The third, cognitive resource, refers to the psychological qualities needed for leadership: intelligence, experience and capacity to withstand stress. Mental capital, in the report's sense, is both a property of the individual and of the nation. For Foresight, 'the idea of capital naturally sparks association with ideas of financial capital and it is both challenging and natural to think of the mind in this way.' The invocation of 'natural', however, carries with it a not-so-subtle evocation of biological inevitability, hence its rhetorical attraction for policy wonks. Who after all can go against 'nature'?

The firm message is that to survive against the emergent Asian economic giants in the knowledge economy of a ruthlessly competitive modern world requires upskilling the labour force, thereby increasing the national sum of mental capital. And just to reinforce the economic thrust of the entire report, *Nature* published a summary reprising Adam Smith, entitled '*The Mental Wealth of Nations*'.[2]

Mental and other capital

Foresight's modifier 'mental' points to those non-material forms that both supporters and critics of capitalism see as vital to understanding post-industrial capitalism. Vital, perhaps, but agreement about the conceptualization remains a matter for continued debate. These formulations are far from politically neutral; thus, where Adam Smith spoke of the wealth of nations, Gary Becker, in his highly influential 1964 book *Human Capital*, focuses on rational individuals and their capacity for purposeful choice. For Becker, as a true marketeer and methodological individualist, such rational self-interest ensures the best for both the individual – by enhancing their personal non-material capital – and the nation. Becker's win–win thesis was influential well beyond the professional circles of economists, significantly informing the educational policies of the West.

Today, it is another Chicagoan economist, James Heckman, whose thinking on education and economic growth holds sway. He, like others, returns to that sixties policy preoccupation with early intervention, symbolized by the US project Headstart. The project was aimed at disadvantaged pre-schoolers, a category in which African-American children were massively over-represented.[3] Heckman draws on the work of the American sociologist J. S. Coleman's *Equality of Opportunity in Education*, which looked to the concepts of social capital and social networks as means of securing social mobility. For Coleman, networks and social capital are neutral, in principle available to all. Amongst the many sociologists who took part in the debates around social capital, the French sociologist Pierre Bourdieu was particularly influential. For him, being in a network stems not from an individual's wish to participate but from their 'habitus' – social location makes a quick if approximate translation. Bourdieu therefore sees networks and social capital as integral to the self-reproduction of elites. For him, education as an institution does not and cannot offer social mobility.

Heckman gives a positive reading to Coleman's work. He agrees that cognitive levels were not enhanced among disadvantaged children through many US pre-school projects, but if followed by high-quality schooling and college education, as in the long-standing High Scope PreSchool Program, there were substantial economic benefits. Research showed that every dollar invested in the programme saved seven in later costs. This one to seven ratio has achieved iconic status for advocates of early intervention. Early and long deep funding for targeted disadvantaged children was the key. Critically, Heckman claims, early preventive intervention is more effective than later remedial interventions attempting to put right what has already gone wrong. To this combination of social and human capital theory, Heckman adds the fast-growing neuroscience of early development to argue that educational policy, beginning with early intervention but by no means stopping there, will both foster social justice and increase economic productivity.[4] A fine claim in the abstract, but it ignores the reality of typical US social policy in which pioneering projects of high quality are frequent, and indeed attract international interest, but are rarely generalized out at the level of an individual state, let alone across the nation.

In its concept of mental capital, the Foresight Report draws on the economists – Heckman and the human capital school – but ignores the sociologists Coleman and Bourdieu. Preoccupied with the individual acquisition of mental capital, the report ignores capital itself, and only weakly acknowledges social and cultural capital. An everyday example of its weakness would be the current political debate about the morality of the advantages some young people have in securing the unpaid internships that open the door to a well-paid and interesting future. These advantages include wealthy parents, social networks such as the parents' occupational, family, friendship and other influential connections, and the cultural capital provided by private education, which give them the confidence and attitudes shared with the providers of the internships. All these the potential internee takes with them into the interview room. These, the unjust mechanisms through which elites self-reproduce, have more in common with the sociologists' theorizing than Foresight's Mental Capital with its roots in the economists' rational-choice theory.

Like Heckman, Foresight's gaze is on the disadvantaged and on how to increase their mental capital, enabling them to contribute to economic growth rather than being a liability on the state. This capital is accumulated primarily during a child's early years, and provides a secure (non-material) bank account that can be drawn on through life to enable cognitive and emotional well-being. For Foresight, genes, developmental disorders, poverty, poor parenting, poor housing, poor schooling limit the mental capital (including well-being as a more modest version of the elusive happiness) that a child might otherwise acquire. Foresight's proposed solutions: early intervention programmes targeting children from poor and deprived social backgrounds, with learning and other disabilities, which would otherwise significantly inhibit their acquisition of mental capital. Prevention is, they argue, more effective than waiting until ill-treated and disadvantaged children are referred to children's services. True, but not exactly cutting-edge policy thinking. In the 1890s, Margaret McMillan, an elected member of the Bradford School Board, shocked by the unwashed and half-naked children of the mill workers, asked: how can a dirty child be educated? And in response campaigned for, and established, nurseries and open-air holidays.[5]

From Foresight to Allen

The innovation of the Foresight Report, and central to the concerns of our book, was its emphasis – and as we shall suggest its over-emphasis – on the importance of neuroscience in developing effective early intervention programmes, not just in connection with already recognized specific learning disabilities such as dyslexia and dyscalculia (discussed in Chapter 4), but as appropriate for all disadvantaged children. Foresight's preoccupation is rather different from that of the neuroscientists themselves, who consider that their insights are universally applicable. The report summarizes the neuroscientific case for the importance of the early years for the acquisition of mental capital thus:

1 Brain mechanisms underpin learning.
2 Most brain development occurs within the first few years of life – at birth, a baby's brain is only 25 per cent of its adult weight, at a year, 60 per cent and 95 per cent by age ten.
3 Neglect, poverty and abuse are stressful and impair the development of cognitive and emotional capacity.
4 Effective interventive strategies directed to at-risk children will facilitate their healthy neurodevelopment.

Therefore (5) interventive programmes require collaboration between neuroscientists, child psychologists, social scientists and educationalists.

These neuroenthusiasms began to influence political thinking. In the same year, 2008, that Foresight was published, the case for early intervention based on neuroscience was pressed in a joint report by Labour MP Graham Allen and Tory Iain Duncan Smith, *Early*

Intervention: Good Parents. Great Kids. Better Citizens. (Theirs was but one of a cluster of similar policy documents, including one chaired by the MP Frank Field and another by social work professor Eileen Munro, stressing the importance of parenting and the need for early intervention, drawing on neuroscientific findings in support.) By 2010, with Duncan Smith installed as Secretary of State for Work and Pensions, Allen was commissioned to update theirs. So keen was he that he produced not one but two reports, published in swift succession in 2011.[6] Both front covers feature MRI images of two brains, the one described as that of a normal three-year-old; the other, much smaller, labelled 'extreme neglect'. We'll come back to these images later in this chapter but, for now, the gist of Allen's argument is a greatly amplified version of the neuroscientific evidence cited by Foresight, that the first three years of a child's life, a period of intense brain development, permanently shape the child's cognitive, social and emotional future, and that proper parenting over these years is crucial. Get things wrong, and the child's brain will not grow properly, with all sorts of negative consequences; get them right, and huge benefits flow. To secure these potential gains, the state should intervene in the interests of the child's mental capital and hence the growth of the economy. The cover of the second

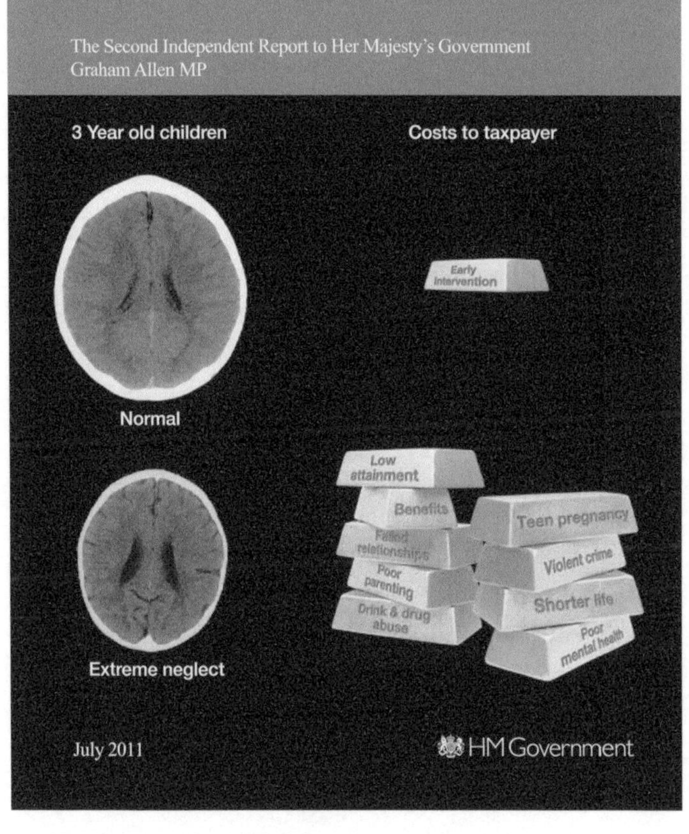

The cover of the Allen Report

report highlights these claims by placing alongside the MRI images a pile of gold bars, variously labelled 'low attainment, benefits, failed relationships, poor parenting, drink and drug abuse, teen pregnancy, violent crime and shorter life' – the heavy costs to the taxpayer of failure to intervene.

The political significance of the redefinition of poverty

Soon after becoming prime minister, David Cameron, in a speech at the Demos think tank, explicitly endorsed Allen and Duncan Smith's approach, emphasizing the importance of parenting over and above poverty for the young child.[7] By the 2015 election, the Conservative Party's manifesto insisted that the root causes of poverty are not a low-wage economy but 'entrenched worklessness, family breakdown, problem debt, and drug and alcohol dependency'. Following the election, the government, now no longer hampered by their Lib-Dem coalition partners, sought to terminate tax benefits but will take until 2020 to raise the minimum wage to a pitiful £9 an hour. With this, they reneged on their own 2010 Child Poverty Act, which had been supported by all parties and which had bound government to end child poverty by 2020. For the new government, the target was 'unsustainable' and in an effort to avoid embarrassment it changed the official definition of child poverty. Hitherto, like every member of the OECD, the UK has defined a child as in relative poverty if living in a family with an income of less than 60 per cent of the average. Now this is no longer the case. Henceforth, poverty is to be defined not so much in terms of relative income, but of educational achievement, worklessness and drug addiction. Remove the innovation of neuroscience, and the combination has much in common with the claimed 'cycle of deprivation' of Sir Keith Joseph, education minister in Margaret Thatcher's government, and the even older antecedent of the nineteenth-century social policy of the moral policing of the poor. Cameron's post-election announcement that child benefits will be withdrawn from families where the parents fail to send their children to school and who have not – regardless of whether they can – paid the fine for non-attendance is of a piece.

'Early Intervention' – Allen uses the capitals to make clear his specific use of the term – fits neatly within this changed concept of poverty. (We follow this capitalization when we are alluding to the Allen/Duncan Smith project and not to the many early interventionist projects which, with or without reference to neuroscience, have long been an element within social policy.) It is to be implemented by social workers, teachers, doctors and nurses specifically trained in methods that will enhance cognitive and emotional development. By his second report, Allen waxes even more euphoric; there will be massive savings to the public-sector finances, fewer prisons will be needed and the structural deficit – a prime concern for the Cameron-Osborne government – could even be eliminated. But the programmes will need resources. And here, in a move guaranteed to appeal to a government committed to privatization and the small state, Allen proposes that his Early Intervention programmes be funded either through the voluntary sector, or through specific outcome-based contracts to private providers. Such privatization and payment by results is a marketized model far removed from Head Start or its UK equivalent, SureStart. Small wonder that his reports were commended by organizations ranging from accountants PriceWaterhouse-Coopers through Portland Capital and Goldman Sachs to the Metropolitan Police. Allen's two reports each ended with the proposal to establish an Early Intervention Foundation and, indeed, by 2013 the Foundation had been recognized as an independent charity with him as its chair.

Allen's belief that the market would play a sustainable role within his Early Intervention projects appears to be based on hope rather than evidence. Certainly the outcome of legislation passed in 2006, encouraging private provision of nurseries – their nearest equivalent – is not encouraging. Only three years later Ofsted reported that, far from expansion, 11,000 nursery places had been lost and 58 per cent of nurseries were still provided by local authorities.[8] The National Audit Office found that half of the private providers made a loss and that only 6 per cent made a surplus.[9] As the chief executive of the British Association for Early Childhood Education pointed out in September 2015, the chancellor's policy of swingeing cuts on social services has led to a 35 per cent budget reduction for local authority children's centres from 2010–11 levels to those of 2014–15.[10] This left local authorities to choose whether they fund children's centres (which support children and their families whose problems substantially correspond to those of the Early Intervention proposals) or services for disabled and older vulnerable people. Unless investors are willing to provide funding on the needed universal scale, with little or no prospect of return, Allen's proposal looks like pie in the sky.

The reports and their neuroscience

For Allen, the first three years are 'far and away the greatest period of growth in the human brain ... synapses in a baby's brain grow 20-fold from having some 10 trillion at birth to 200 trillion at age 3'. It is the period over which a baby attaches to its mother, and the secure environment that this provides is the substrate for full cognitive and emotional development. The Allen Reports were followed by a further proposal for early intervention, *The 1001 Critical Days*, sponsored by the Wave Trust (a foundation dedicated to tackling the roots of violence) and the National Society for the Prevention of Cruelty to Children, with an introduction by the government's chief medical officer, Sally Davies, and endorsed by a cross-party group of MPs, ranging from the Tory Andrea Leadsom to the Green's Caroline Lucas.[11] The '1,001 Critical Days' moves Allen's timeframe back to conception, recognizing that the nutritional status of the woman at conception has powerful predictive value for the health and well-being of the child.

However, for governments to will the resources to improve the nutritional well-being of the nation requires political commitment or a potentially overwhelming crisis like nation-state war. Facing acute shortages in food supplies in the Second World War, Britain produced a food plan based on the nutritional needs of the people, with priority in rationing for pregnant women, children and those doing heavy manual work. Infant mortality levels declined, working-class children became taller and civilian life expectancy, despite the bombing, increased. The food insecurity resulting from the economic crisis produced by the banking failure has, however, not generated a similar political response. One in every four children is living in poverty – 1.7 million in severe poverty. While accurate statistics are difficult to determine, the Trussell Trust estimated in 2015 that a million Britons would soon be reliant on food banks.

'Why', the 1,001 Report begins, 'is the Conception to Age 2 period so critical?', before going on to marvel at the brain's rapid growth rate: 'By the 1001st day the brain has reached 80% of its adult weight .... *From birth to age eighteen months*, connections in the brain are created at a rate of *one million per second*!' (their emphases). Even more than Allen, the 1,001 Days Manifesto emphasizes that this is the period in which attachment, 'the bond between a baby and its caregiver/s is formed'. Drawing on a wealth of social rather than neuroscientific research, they emphasize that a baby's social and emotional development depends on the quality of their attachment to their primary caregivers. Further, 'a foetus or baby exposed to toxic stress (and its biochemical surrogate, the hormone cortisol)' can have their responses to stress distorted in later life. Such stress arises if the mother is suffering from 'depression or anxiety ... a bad relationship ... bereavement'. The implications are clear; as the 1,001 Days Manifesto spells out, 'Ensuring that the brain achieves its optimum development and nurturing during this peak period of growth is therefore vitally important, and enables babies to achieve the best start in life.' If the mother/caregiver cannot/does not give such support during this critical period, the damage will be virtually irreversible.

All the elements of the appeal to neuroscience are thus in place – critical periods, brain growth, synapse number, stress, cortisol levels – plus attachment theory with its focus on the relationship between the child and the primary caregiver. The appeal is certainly there; the question is whether the neuroscience supports it.

The origins of those MRI images of normality and neglect

But we should begin with those dramatic MRI images on the covers of the Allen Reports and extensively replicated, comparing a 'normal' three-year-old versus one suffering from 'extreme neglect'. The report quotes the source of these as a paper by Bruce Perry of the Child Trauma Academy in Houston, Texas, published in a short-lived journal, *Brain and Mind*.[12] Seek the origins of the images in *Brain and Mind* and one is referred back to a poster presentation (therefore unrefereed) by Perry at the 1997 meeting of the American Society for Neuroscience.[13]

The *Brain and Mind* paper describes 122 neglected children referred to Perry's Academy clinic. He classified the children into four groups: 'global neglect' (relative sensory deprivation or lack of social interaction), chaotic neglect, chaotic neglect plus drug exposure and global neglect plus drug exposure. Forty-three of the 122 children received MRI brain scans. Of these, twenty-six were classified as having experienced 'chaotic neglect with or without drugs', and three were reported as showing abnormalities. Of a further seventeen classified as having experienced global neglect, eleven were reported as showing abnormal brain scans. There is no further information on these children or their brain scans; even the routine categories of age and sex are missing. The images Allen uses are the only ones in the Perry paper, and they show such an extreme difference – far more dramatic, for example, than anything seen amongst the desperately impoverished children rescued from Romanian orphanages in the 1990s following the fall of the Ceausescu regime[14] – as to make further questions regarding their origin imperative.

So we wrote to Dr Perry for this information. He replied that they had not published this 'initial set of observations' and that it had 'become clear that due to a variety of factors, we were really unable to conclude much more than "severe neglect impacts brain development." The tremendous variability in the nature, timing, pattern of abuse/neglect . . . led to such a heterogeneous sample' that they needed to develop methods to 'better interpret any biomarker or neuroimaging data'.[15] To distance himself from the Allen Report, he claimed that Iain Duncan Smith had 'distorted' his research into childhood neglect.[16]

The Allen Report with its MRI brain images of the severely neglected child was widely distributed to public-health officers, shocking and convincing many, although others remained sceptical. Perry's influence and his MRI brains have entered UK private-sector training programmes such as the Solihull Approach, which offers early intervention-based courses for social workers, health visitors, nurses and parents. The MRI 'normal' and 'extreme-neglect' brains were also used in an advertising campaign by Camila Batmanghelidjh's now-defunct Kids Company, closed down abruptly in 2015, having allegedly mismanaged some £46 million of public money, some directly allocated by successive prime ministers over-ruling their own civil servants. The Kids Company ads featuring black teenagers and the Perry brains were banned in 2009 by the Advertising Standards Authority, on the grounds that they were both racist and implied that brain size was related to emotional neglect and violent behaviour – for which, the regulator said, there was no evidence.[17]

Neuroscience, development and early intervention

What of the neuroscience behind Allen's and the 1,001 Days reports? Before looking at their claims in more detail, we look briefly at how neuropsychologists see the normal pattern of infant development, though a word of caution is required. In this biological narrative, childhood and adolescence are universal biological entities, categories independent of class, gender, ethnicity and geography. For those committed to examining the social and the historical, matters are more complex. Historians have explored the history of childhood; just when and where was the concept first used, when did specific social practices begin, such as being dressed with clothes designed for them rather than with small-scale adult clothing? And when were they first seen as a potential new market? The concept of the teenager did not exist in 1900; tentatively used in the US in the interwar period, by the 1940s the teenager with distinctive dress and habits was a recognized social category in the US, with the UK following in the 1950s. The extension of the school-leaving age is widely recognized as the main cause. The more conservative French language still has no such word; adolescence and *les jeunes* are sufficient. Given the proliferation of new identities within Western culture, such as the recent appearance of the 'pre-teens', the historians' work gnaws at the neurobiologists' positivistic certainty concerning such phenomena as 'teen sleep', which we discuss in the next chapter.

Developmental biology documents how newborn babies emerge from the relatively tranquil and ordered environment of their mother's womb into a world rich with environmental and social stimuli, sounds, smells and a chaos of visual impressions. They need to engage not only with a radically different physical environment from that in the womb, but also to negotiate a complex social world. Along with motor (grasping, reaching) and cognitive skills, the baby must develop social skills – neuroscientists have come to speak of the 'social brain' (somewhat hopefully, as their research methods still focus primarily on the individual or the dyad of mother/ baby and are still at an early stage), which develops over the years as the brain matures and different regions come on line over the first decades of life.

Even at birth, babies are primed to engage actively, to extract what is important about their environment, learn and act upon it. One of the easiest ways to study this is to observe a baby's gaze. Given a choice, newborns will prefer to look at faces rather than random arrays of dots. And within weeks they will preferentially respond to their mother's/primary caregiver's face, rather than a stranger's. Ingenious experimental techniques make it possible to identify the brain processes and regions involved in these responses non-interventively. Babies can be fitted with 'hairnets' studded with sensors, able to pick up minute changes in the brain's electric activity as they shift their gaze or attention, without interfering with the child's activities or mood. The technology of the 'hairnet' thus enables developmental neuropsychologists to relate a baby's actions to brain processes. A typical experiment might compare the electrical record when baby and mother are looking at one another with that when the mother turns aside. When this happens the baby may show distress, and this will be reflected in the electrical signal. Such studies show that the brain region in the right hemisphere (inferior temporal gyrus) activated by their mother's face in two-month-old babies matches the 'face region' in the adult brain – a finding heavily drawn on by the attachment theorists.

By seven months, infants can discriminate between happy and fearful – though not fearful and angry – faces and speech; that is, they can perceive emotions, and by eight months can follow another person's actions, such as turning away to look at a nearby object. The brain processes involved in all of these, considered to indicate the development of the baby's 'social brain', can be observed via the hairnet. How do the laboratory observations match the assumptions of the Early Interventionists? As they emphasize, a baby is born with a lot of brain growth still to do, and much of this post-natal increase reflects the growth of synaptic connections as the brain wires itself up in an unrolling developmental sequence, with different regions growing at different rates as they 'come on line'.

The Early Intervention assumptions are that: (1) the more synapses the better; (2) poor environments in these critical years permanently reduce synapse number and the brain doesn't wire up properly; (3) there are critical (or sensitive) periods in brain development, especially; (4), for proper attachment bonds to be formed between caregiver (mother) and infant; (5) 'toxic stress' at this early period has lasting consequences for later development.

Neither of the first two is supported by the neuroscientific evidence; the third, fourth and fifth oversimplify very complex relationships between the child's developing brain and their social and environmental context. The neurocentrism of both these Early Intervention proposals fails to take account of Foresight's requirement for collaboration between neuroscientists, child psychologists, social scientists and educationalists. So let's explore the claims and the science.

(1) Synapses – the more the better?

According to one of the early interventionist training programmes drawing strongly on Perry, 'an adverse environment will lead to a child having 25% less synapses or connections in their brains than they could have had, while a stimulating environment can lead to 25% more connections'.[18] One of the key features of the early development of the brain is a vast overproduction of neurons and synapses. Just as it takes myriads of sperm to ensure that one is able to reach and fertilize the egg, so it takes myriads of neurons to be born during development to ensure that some survive and wire up appropriately. A process called apoptosis – programmed cell death – then removes the surplus. So too with synapses, whose proliferating numbers during early development are steadily pruned away, so that by adulthood in some brain regions there are less than half of those present at age three. (These anatomical measurements can only be made post-mortem; there is no way of counting cells or synapses in the living human brain.) Neuroscientists believe that this pruning of redundant synapses both removes unused connections and improves the efficiency of those that remain. Another problem with the synapse number claim is that it implies that once a synapse is made, it remains in place. However, time-lapse studies of the brain in experimental animals shows that synapses are highly dynamic, continually being modified, disappearing and being reformed throughout life. ('Use it or lose it' is no bad slogan.) Indeed, this remodelling capacity – plasticity – is the neural mechanism that enables a person to learn from experience, remember and change how they respond. Every thought or action leaves its trace in the brain. So the claim about the significance of synapse numbers is misleading.

(2) Enriched and impoverished environments

Humans are unique amongst our closest evolutionary relatives in the extent to which our children are born prematurely, and do much of their development postnatally (the increase in brain size over the first years of a child's life is largely due to an increase in connections between neurons). However, the foetus's pre-natal environment is important. Most neurons, along with crucial neural pathways and connections, are in place while the foetus is still in the womb, long before birth. Their orderly development can be affected by maternal well-being. Stress, poor nutrition and, even worse, starvation, especially around the time of conception, also have long-term effects on a child's intellectual and physical development, as studies of survivors from the famines in Rotterdam and Leningrad during the Second World War have shown.[19] Further studies of the descendants of the Rotterdam children have shown that some of these negative effects carry forward into the next generation too. (This finding surprised many geneticists; part of the explanation seems to lie in epigenetic processes – the ways in which genes can be lastingly modified during development.)

Are synapse numbers and brain structures modified by environmental experience postnatally? The evidence from animal studies is clearly yes. The Early Intervention literature regularly refers to experiments dating back initially to the 1950s, on the effects of deprivation on brain development in rats. Rats reared in so-called 'impoverished' environments, kept in isolation in bare cages, have thinner brain cortices and fewer synapses than their littermates in 'enriched' environments – reared with companions and toys to play with, in larger cages. The problem with extrapolating these results to humans, however, is that even the rats' 'enriched' environment is impoverished by comparison with the busy, smelly world in which wild rats thrive, let alone the complex physical and social environments that most growing children experience.

Less frequently commented on in the Early Intervention literature is that even in adulthood, rats reared in the impoverished environment then transferred to an enriched one show considerable catch-up. Is this relevant to humans? Follow-up studies of the Romanian orphans who had been adopted into other European homes found that most were able to recover to something close to a normal trajectory in both brain growth and behavioural outcomes.[20] Even following early malnutrition, which slows brain and body development, children can recover; given improved conditions, a growth spurt enables considerable catch-up.

But this is an optimistic account. It must be balanced against a pioneering example of collaboration between social scientists and neuroscientists in the US, which examined the development of children's brains in relation to socioeconomic status, published in 2015.[21] The team studied 1,099 'typically developing' youngsters between the ages of three and twenty, and found that the surface area of the brain was related to family income. Amongst poorer families, a small increase in income increased brain area significantly – especially in brain areas associated with language and reading skills. If the family was rich, then an increase in income made little difference. The implication is that the simplest and most effective of early interventions to increase mental capital would be to lift children out of poverty. What has been happening in UK policy since the election of the Conservative Party, initially as a coalition government in 2010, is precisely the reverse: more children are being pushed into poverty – as measured by access to resources – while the government officially abandons the pledge to lift all out of poverty by 2020.

(3) Sensitive periods

Back in the early days of the field, neuroscientists drew on the work of Konrad Lorenz in the 1930s, on imprinting in young goslings, to suggest there were certain critical periods in brain development, such that if the appropriate stimuli did not occur during this time window, the damage could not be rectified afterwards. However, neuroscientists today reject the term 'critical' as too absolutist and determinist, and speak of more flexible 'sensitive' periods. One example of such sensitive periods in humans comes from the 1960s, with studies on squint, a condition in which a person's eyes look in different directions. In the UK, about 5 per cent of babies are born with a squint; in many cases this is self-correcting within a few months of birth, but if it does not correct, it may become permanent. Treatment involves putting a patch over the good eye for several hours a day, so that the child – assuming they are willing to keep the patch on – uses the affected eye. But the treatment won't work in children older than seven; the neural pathways have become fixed. Similarly, there seemed to be a sensitive period for learning to distinguish phonemes. It is well known that adults who have been brought up hearing only Japanese can't distinguish the sounds of 'l' and 'r', which have no equivalents in the Japanese linguistic system. But Japanese children brought up in a European-speaking environment during the first six months or so have no such difficulty.

Two interacting brain processes are involved in such developmental sequences: specificity and plasticity. Specificity ensures that, during normal development, the correct pathways wire up, such that different brain regions are properly connected and remain so even during periods of rapid growth. A good example is the connection between eye and brain. The optic nerve runs from the retina via a brain region called the lateral geniculate to the visual cortex at the back of the brain. During development, eye, lateral geniculate and cortex all grow, but at different rates, meaning that the synaptic connections between them must continually be broken and then remade, but without impairing or disrupting vision. This sequence must be relatively impervious to the growing child's experience or environment. But the key word here is 'relatively', as the treatment for squint shows. This is where plasticity comes in, as experience fine-tunes the connections; thus, after a period of early receptivity to faces in general, babies learn to distinguish their primary caregiver's face from others.

The term 'plasticity', though, has many meanings. In the sense of the previous paragraph, it describes one of the necessary processes of development. But it is also used to describe the – limited – capacity of the brain to repair itself after injury, as well as the delicate remodelling of synapses that accompanies learning, that making and breaking of connections described earlier. Early interventionists often use the term as if it described some new discovery about how the brain works, and it has become a rhetorical device with which to mobilize support for the potential of such interventions. But plasticity as neuroscientists understand it is neither unlimited nor necessarily positive in its capacity to remould the brain; rather, it is part of the rich and varied dynamic processes through which all living organisms interact with their environments throughout their life cycle.

(4) Stress and cortisol

Stress is notoriously difficult to define. A little is necessary to respond effectively to the daily challenges of life; too much over too long a period can leave some unable to act at all. And there are huge differences in individual responses. What is a good and helpful level of stress for one person may be debilitating for another. It's been known since the 1950s from experiments with rodents and monkeys that acute stress in infancy, including what was described as deprivation of maternal care, can have lasting physiological and biochemical consequences, affecting resilience and susceptibility to disease in later life. How far the effects of these extreme experiments can be extended to humans, with our extraordinary capacity for plasticity, of transcending such seeming determinism, is uncertain.

The difficulty with these early interventionist projects, despite their expressed concern for children's well-being, lies in their rhetorical appeal to neuroscience as providing biological legitimacy. Thus, a key component of Allen, 1,001 Days and related programmes is the link between stress and a hormone, cortisol, secreted by the adrenal glands which are located just above the kidneys. Cortisol has multiple effects across the body, from regulating blood sugar, salt and water balance to learning and memory. Blood levels of cortisol vary through the day, being highest in the morning and lowest at night, and also across the life cycle from infancy to old age. Furthermore, the levels are quite labile: stress, from the need to meet a sudden challenge, to chronic anxiety and life hazards, all increase cortisol levels at least briefly. All this means that any single measurement of the blood level of the hormone is not very informative because of the daily variation, but it's been found that it accumulates in hair. Hair grows at a centimetre a month, so the cortisol content of that centimetre could be seen as an index of the level of a person's stress over that month. So some of the early intervention protocols propose routine sampling of a baby's hair to provide an index of chronic stress, thus providing a biomarker for 'toxic stress'. However, because there are large differences in cortisol levels between individuals – 'baseline levels' measured mid-morning may vary fivefold between one person and another, and across different populations – a direct correlation between cortisol in hair and stress levels is hard to make. Exposure to the blue light emitted by electronic devices like smartphones and computers, especially late at night, affects cortisol levels and intellectual functioning the following day – a self-inflicted exposure by some and an occupational health risk for others. Very high levels are associated with endocrine diseases such as Cushing's syndrome, while very low levels are linked to other diseases.

Early interventionists in general – not just Allen and Duncan Smith – tend to ignore such complexities. They set aside individual differences, instead asserting that high cortisol levels are indicative that an infant has been subject to 'toxic' stress as a result of an unsupportive environment. One popular book on the reading list for such programmes, Sue Gerhardt's *Why Love Matters*, goes so far as to refer to it as 'corrosive cortisol'.[22] Measuring cortisol levels as an index of stress is not dissimilar to looking for a lost key under a street light because nothing can be seen in the darkness.

(5) Attachment

Attachment theory was developed in the 1950s by the psychiatrist and psychoanalyst John Bowlby to emphasize the importance of the strong affectual bond between mother and baby for the healthy psychological development of the child. Having worked with deprived and traumatized children, Bowlby became interested in the observational research methods of the ethologist Robert Hinde. Convinced of the potential of the method, he and his colleague Mary Ainsworth carried out a number of field studies of infants and their mothers. This move separated him from mainstream child analytic theory, then more interested in the psychic life of the child rather than their lived experience.

He drew on ethological research on the long-term effects of separating baby rhesus monkeys from their mothers to propose that a newborn baby was programmed to form such an attachment over the first months of life. Despite Bowlby's interest in ethology, the psychoanalyst forgot the famous photograph of Konrad Lorenz's new-hatched goslings closely following his wellies, as happily imprinted on them as they usually are on their parent. Odd, if transitory, relationships are not that uncommon among animals, as in the video clip of ducklings cuddling up to a cat, elbowing her kittens aside to forage as ducklings do, thus securing the milk intended for her kittens. But boots as significant others should surely give attachment theorists pause for thought. In developing his thesis, Bowlby was over-influenced by his reading of the ethologists, even though they did not study humans – and probably by his own lonely childhood. As the son of an upper-middle-class family he had, as normal for that stratum, been cared for by a nanny and sent away to boarding school at eight, bitterly resenting being separated from both his nanny mother figure and his mother.

Attachment theory, though well received by many, was sharply criticized for both the quality of the empirical data and the woolliness of the attachment concept. As the distinguished social scientist Barbara Wootton tersely observed, all it means is that, like adults, children need love. Under pressure, Bowlby and Ainsworth found themselves adjusting the model, first admitting fathers, then introducing the more neutral concept of primary carer, rather than just biological mothers. Although today the language of mother/primary caregiver is extensively used, as by Gerhardt, attachment theory was born in the 1950s context of the idealized nuclear family, the employed father and the full-time mother and housewife. In the twenty-first century few families correspond to this imaginary; parents no longer feel it obligatory to marry or to be in a heterosexual relationship to have children, while lone parenting is common. Partnerships break and new ones form; there are more intercultural families, and what is increasingly named the new blended family has arrived, bringing not just second fathers or mothers but also ready-made new siblings. And with the increase in assisted conception, historically new family forms are brought into existence. But in all this melee of change, it remains that it is predominantly the mother who does most of or coordinates the care work, with grandparents doing childcare while parents work, particularly in school holidays.

However, while these changing patterns of family organization have been extensively studied by sociologists, social psychologists and anthropologists, it wasn't until the primatologist Sarah Hrdy's 2009 book, *Mothers and Others*, that the dominant ethological narrative drawn on by Bowlby was challenged from within. Hrdy compared the infant rearing practices of other primates such as chimpanzees with those of women. A key difference was that where chimp mothers rear their infants alone, allowing only their own genetically related kin to hold them, human mothers are willing to let trusted unrelated others hold and interact closely with their babies and children. Her term for this is 'alloparenting', or 'parenting by others'. But in practice what human parents do is hugely determined by their access to resources. Thus, alloparenting was made universally available in the Nordic welfare states with their provision of free nurseries and the expectation that women would be employed, made possible by a high level of taxation. Such a situation never occurred in the US, with its distaste for collective provision of childcare, nor in patriarchal Britain. Hrdy's concept of alloparenting works better with the UK wealthy, who once again employ nannies and send their children to boarding school (an ironic return to Bowlby's own formation). If they are lucky, parents send their under-fives to public nurseries and to fee-paying ones if they are not. Low-income families are provided a specific number of hours of free childcare for very young children. Childcare today does not easily fall under a single unifying concept; it is much more a patchwork quilt.

Attachment theorists in recent years have avoided the perils of Bowlby's too close embrace of ethology, but at the cost of missing out on the relevance of Hrdy's work in reducing the importance of a single primary caregiver and her genetic relatives in favour of alloparenting. Instead they have looked to the neurosciences for support, seeking to root attachment theory into brain development, noting that the seven-month time course during which attachment is supposed to be formed parallels that of the maturation of right-brain systems associated with affect and self-regulation. (The early intervention literature is full of claims about faulty connections between the 'cognitive' left brain and the 'emotional' right brain, a 'neuromyth' that modern neuroscience regards as much oversimplified, as we discuss in Chapter 4.)

The problem with the attempt to link attachment to brain development is similar to that we discussed earlier about the attempts made by psychoanalysts to 'neuro-ize' their theories. 'Attachment' comes from a different disciplinary vocabulary from 'right brain'; translating one to the other is fraught with difficulty. Attachment, according to researchers observing mother–baby interactions, is not a sudden event like turning on a light but a development over several months; there are no biomarkers to indicate that a baby is or is not attached. What is left is attachment theorists' deference to neuroscience's account of the developing brain, while excluding the multitude of other brain and bodily changes which occur over these months and which cannot be teased apart. There may be no harm in this piece of neurospeculation, except when it is heavily over-interpreted, as in the programmes and publications of the early interventionists, to argue that unless mothers interact in a prescribed way with their infants over these seven months irretrievable damage will be done to the child – attachment bonds cannot form and the child will grow up emotionally deprived. Nothing like guilt-tripping the almost certainly over-stretched mother, much as unreconstructed Bowlbyism was mobilized against employed mothers in the fifties and sixties.

It is a step away from our main preoccupation with neuroscience, but it is worth pointing to a similar instance of how children have continued to need protection from over-extrapolation of biological findings. The heady mix of animal studies, speculative evolutionary psychology and criminal statistics, drawn on to explain the sexual and violent abuse of children, is a classic example. Researchers Martin Daly and Margo

Wilson claimed that a man – whom they speak of routinely as a stepfather – living with a woman but who has no genetic connection with her children poses a risk to their safety through sexual abuse, violence and even murder, rather as a new male lion taking over the pride usually kills the cubs fathered by his predecessor.[23] While there was extensive academic criticism of both their theory and their research methods, it took the courage of the sexually abused children, now adults, to expose their abuse, not by their non-genetic fathers, but by their Father the Catholic priest. Collectively, the victims' exposure – despite powerful resistance – of the institutionally tolerated paedophilia of Catholic priests and that of other paedophiles with easy access to children drove the Daly and Wilson thesis out of the media and out of public view. If it lives on, it will only be in some obscure academic ghetto, unlikely to cause harm. Our point here is that bad science can fail the vulnerable and distract attention from the real perpetrators.

The bridge is still too far

Nearly twenty years ago, in response to the burgeoning claims of neuroscience to direct, or at least advise, on childrearing and educational practice, the American philosopher John Bruer wrote a major critique, arguing[24] that while research in cognitive psychology was already contributing to understanding children's intellectual development, and while neuroscience could illuminate cognitive psychology, to make the leap directly from neuroscience to childrearing was 'a bridge too far'. His book, *The Myth of the First Three Years*, critiquing the ways in which the early interventionists draw on the neuroscience, is as relevant now, even after the subsequent dramatic growth of neuroscientific knowledge, as it was in 1999, though entirely ignored by the many experts quoted in the Foresight Report with its efforts to, as they put it, make 'the most of ourselves in the twenty-first century'.

That report was published just before the banking crash that ushered in an era of policies of austerity as successive governments baled out the bankers by slashing welfare. In this new milieu 'making the most of ourselves' has, for the poor and the poorest, become a mere ironic trope. In 2015, when a now entirely Conservative government redefined poverty not by relative income but by educational achievement, worklessness and drug addiction, targeted intervention/moral policing came once more to the forefront of policymaking. Duncan Smith, one of its principal advocates, now Secretary of State for Work and Pensions, is on record as believing that early intervention for the children of failing parents can avert the huge costs of inaction evoked by the gold bars on the cover of Allen's report. With this emphasis, it seems probable that more commercial operations such as the Perry-influenced Solihull Approach, with its training programmes for social workers and parents, will both increase in numbers and position themselves to secure those outsourced contracts that Allen plausibly if unrealistically believes will come into existence at minimal cost to the taxpayer.

The doom-laden ideology of early interventionism has learned nothing from Bruer. They continue to imply that a child's destiny is fixed within these first three years of brain growth and synaptic proliferation, and that poor parenting results in the child's failure to form secure attachment, with dire consequences both for the child and economic growth. The alarmist claims about rates of brain growth, synapse numbers, sensitive periods and cortisol levels are at best still a bridge too far, at worst reliant on ideologically driven, bad or over-interpreted science.

Above all, they ignore the consequences of the growing inequalities of a society of the 1 per cent obscenely rich and the 99 per cent of the rest.

Foresight recognizes the costs of inequality to well-being but seeks, following Heckman, to mitigate this through education. Missing is any recognition of the structural links between a globalized capitalism and intensifying inequality. Medical statisticians Richard Wilkinson and Kate Pickett's 2009 book *The Spirit Level* demonstrates the socially destructive impact of such inequality on health and well-being. The impact is not confined to the poor and the poorest, where there is no certainty for parents whether they can put a meal on the table or roof over their heads for their children, or for older people to heat their homes, but to the very fabric of society. Furthermore, as the precariat expands, and the once secure middle classes experience a new insecurity of intermittent unemployment, the housing crisis and an intensifying austerity policy, there is more ill health, both physical and mental, more violence and more abuse of children. Where individual childen are selected as requiring free school meals they perform less well, from the acquisition of early skills to their later grades in GCSE. Research reports that these selected children feel stigmatized and frequently experience bullying. Universal free school meals, as introduced by some local authorities, avoid this stigmatization. A survey of children receiving free school meals reported that they felt stigmatized, excluded and frequently experienced bullying. Inequality, as Wilkinson and Pickett insist, carries severe and widespread social costs which can only be met by structural reform, about which neuroscience has nothing to say.

These issues are ones we will return to in the final chapter. For now, it is time to turn from policies drawing on neuroscience in targeting disadvantaged babies and pre-schoolers, to proposals to put education on to a neuroscientific foundation, and so to change the minds of those being educated.

FOUR

Neuroscience Goes to School

A booming industry

According to the OECD, in a report issued in 2007, 'Today, it is useful, even essential, for educators and anyone else concerned with education to gain an understanding of the scientific basis of learning processes.'[1] The OECD is not alone in this view of the need for teachers to understand the brain. It was followed in 2008 by the Foresight Report we discussed in Chapter 3. A few years later, the Royal Society joined the chorus with *Neuroscience: Implications for Education and Lifelong Learning*. In 2015, *Nature* launched a new journal, *Science of Learning*, whose opening editorial claimed 'we are in exciting times for neuroscience, where the merger of neuroscience with education takes us from the molecular and cellular understanding of brain function to the classroom'. Endorsed by such authoritative bodies, the study of the brain has arrived at the centre of education policy. As a keyword in Google, educational neuroscience generates over 34 million hits.

This is a boom industry. There is an active MBE (Mind Brain Education) Society and a raft of academic research posts and professional journals. It has also offered a market opportunity. Even a decade ago, schoolteachers were receiving up to seventy mailshots a year from companies offering products and courses in 'brain-based education', according to the director of the Cambridge Centre for Neuroscience in Education, Usha Goswami.[2] Responding to a Wellcome Trust questionnaire,[3] 88 per cent of the Trust's panel of science teachers thought that over the next decade neuroscience would improve teaching, and were keen to collaborate with researchers given the opportunity. Admittedly, the teachers had become panel members because they had responded to Wellcome's invitation; nothing wrong with that in itself, but it would be startling if this self-selected group was anything other than interested and positive about the neurosciences' potential contribution to education. Wellcome's 88 per cent is unfortunately all too likely to be used as if it was a finding from a methodically robust study. In this way, op. ed. columnists and the political classes all too easily forget the caveats, and the percentages quietly become set in stone to be cited as unchallengeable evidence. No wonder the neuromarketeers are confident they are pushing at an open school gate.

What is disturbing about the entire neuroeducation project as proposed by the reports is that the learning brain they describe is curiously disembodied; the learning child is replaced by a free-floating organ. *Nature's* new journal takes this further with its editorial statement, which leaps directly from molecules and cells to the classroom. These neuroeducators seem to have forgotten biomedical science's history of that long expensive gap between the lab bench and the patient's bedside, which can fail even at the final trial. They seem not to recognize that to translate neuroscience's research findings to the classroom is likely to be more rather than less complex than translating a lab observation into a new and successful drug. The classroom – the pupils and their teacher – is not just the sum of individuals but a complex social system in a process of constant subtle change. Nor is there an unknown country of classrooms waiting to be discovered by the neuroscientists. Research residents from the humanities and social sciences have been there for a long time and have established a substantial research base, which underpins educational theory and practice. For them, the vogue for neuroization risks being seen as an act of cultural imperialism.

Enhancing educational achievement

According to a 2010 OECD report on the economics of educational achievement, it is the standard of school maths that predicts economic growth and hence wealth.[4] The report estimates that, if the UK had improved the standard of the 11 per cent of children who failed to make the international educational standard minimum between 1960 and 2010, there would have been a 0.44 per cent increase in GDP, and it adds that learners with poor maths are more likely to find themselves unemployed, depressed and in trouble with the law.[5] While statisticians have long been concerned about the validity of the OECD's educational ranking system, PISA (Programme for International Student Assessment), as different tests were used in different countries with some cultures being advantaged and samples sometimes being too small, governments worldwide have nonetheless chosen to use them uncritically to justify sweeping education reforms. (This is not to argue that PISA is useless – but that the data needs handling with caution.) Michael Gove, the then Secretary of State for Education, was no exception. Ignoring the criticism that the number of British schools taking part was too small to produce reliable data, he steamed on, invoking the UK's poor PISA rating to embark on a series of controversial reforms. His abuse of the statistics of the PISA ratings led to his formal criticism by the UK Statistical Authority. This made little impact on Gove. Seeing that Shanghai was top of the PISA league, he decided the Shanghai model must be followed.

Hence primary schoolteachers were sent to Shanghai to study first hand the classroom maths teaching at their experimental primary schools, and a group of thirty specialist Shanghai teachers recruited on two-year contracts to teach maths in English primary schools at a cost of £11 million. Evaluation research reported a small but statistically insignificant improvement. Yet, before committing this not insubstantial sum, it would perhaps have been worth looking at the greater number of hours per year studied by children in Chinese schools compared with those of the UK, and the level and depth of the teachers' professional education. Letters from schoolteachers, commenting on the *Guardian* report on this initiative, pointed out that UK primaries rarely have specialist maths teachers; most teach across the full range of subjects, whereas the recruited Chinese teachers had five years' university training specializing in maths education and typically taught only two lessons per day, with the rest of their timetable devoted to working with colleagues to improve teaching and learning methods.[6] By contrast, the UK government has permitted unqualified staff to teach in state, academy and 'free' schools. In 2014, no fewer than 400,000 children were being taught by unqualified teachers.[7]

By 2015, the government was sufficiently concerned about the shortfall in maths, science and engineering teachers that a £24 million programme was launched to upskill 15,000 non-specialist teachers in maths and physics[8] – around £1,600 a head. While launched with the usual overblown language, when the figures are broken down, this sum would pay for around four to six weeks' tuition per teacher at current university-fee levels, a pathetic sum if the task is to upskill those who are to become the nation's maths teachers. Having already paid some £11 million to bring in the thirty Shanghai expert maths teachers, it would seem that the minister has need of urgent upskilling in arithmetic, let alone maths. Contrasting this £35 million with the £700 million a year subsidy enjoyed by public schools such as Eton and Westminster, with charitable status, provides a conspicuous example of the Matthew principle: 'To him who hath, more shall be given.'

So, in this unpromising context, what chance do the neurosciences have of offering guidance towards identifying teaching practices to help raise standards in general, and aid children with specific learning needs in particular? Neuroscience's advice on how best to teach is unable to take account of the socioeconomic causes for a student's lack of mental capital: under-financed state schools, increasing numbers of teachers leaving teaching after ten years, ever increasing numbers of children in poverty, 46 per cent of children in some primary schools having other than English as their mother tongue, and the correlation between family income and A levels remaining unbroken. Newspaper pictures of schoolchildren jumping with joy at their A and A* grades is no substitute. Blaming the children, their parents and the schools for the lack of aspiration is a popular habit among the political classes but grossly over-simplifies the research findings. Thus, one study comparing ethnically diverse and white working-class areas suggests that the ethnically diverse had both high ambitions and high performance, whereas young people in the latter group aimed at more traditional jobs. As a major report from the Rowntree Foundation states: 'Policy to increase social mobility needs to go beyond assumptions about aspiration – it needs to tackle barriers to fulfilling them.'[9]

More positively, neuroscience has been able to contribute to understanding the developmental learning disabilities of children such as those with dyslexia and dyscalculia. The biochemical and physiological processes that encode new information within the plastic brain are active research areas, and educational neuroscientists hope that unravelling those processes could point to more effective forms of classroom teaching. However, in the classroom, a diagnosis backed by neuroscience can be a two-sided weapon; thus educational research studies have documented the downside of diagnosing/labelling children as having a biologically based learning disability. Teachers may give up on believing the child to be educable.[10]

Enhancing the brain

The marketing of drugs and electrical devices to enhance learning offers individual students the prospect of positional advantage in competitive situations, thus finding both a comfortable and profitable niche in the booming brain-optimization market. An invisible alliance develops between human-capital enhancers, including educational policymakers, neuroeducationalists, producers and sellers of both soft and hard educational technologies, together with all those seeking to enhance their mental capital. These range from the student who takes Ritalin or Modafinil to get better grades, to the pregnant woman who, following the suggestion of one neuroscientist, plays Mozart to expand the brain of the foetus she carries. Conceptualizing the brain as capital, a resource to be expanded, transforms who – or what – is to be expanded. It separates the part from the whole: now, instead of it being the learning child, the student, the ambitious trader, the elder fearing dementia, it appears from some of the titles on the expanding neuroshelf that it is the learning brain, the social brain, the emotional brain, the ethical or the telltale brain which is to be enhanced.

Attempts to enhance human potential, performance and pleasure are age-old. But what is new is neuroscience's very specific focus on cognitive enhancement by way of chemicals specifically targeted at underlying neural processes, primarily neurotransmitters. One of the earliest was amphetamine, initially marketed as a decongestant, Benzedrine, by Smith Kline French in the US in the 1930s (and still marketed as such in the 1950s), and extensively used, along with its close and more toxic relative, methamphetamine (crystal meth) as a stimulant by both Allied and German forces during the Second World War. Amphetamine's addictive effects led postwar governments to make it a prescription drug, but it was nonetheless readily available as a street drug. Amphetamine is still used by combat pilots on long missions, though since the Gulf Wars it has largely been replaced by Modafinil, a prescription drug originally developed as a treatment for the sleeping illness narcolepsy. Illicit crystal meth remains a drug of choice for its euphoric and aphrodisiac effects, despite the savage personal and social costs.

In the 1960s, a close chemical relative of methamphetamine, methylphenidate (Ritalin), began to be prescribed in the US for the treatment of what was then called Minimal Brain Dysfunction (now renamed Attention Deficit Hyperactivity Disorder or ADHD). Amphetamine itself was relaunched as Adderall in the 1990s, as an alternative longer-lasting drug for ADHD. Both are extensively prescribed, and readily obtainable without prescription over the internet, or traded in the school playground. Although known collectively as smart drugs, Ritalin and its relatives do not directly affect cognition itself but improve attention by interacting with the dopamine neurotransmitter system in the brain, making it easier for students – whether or not they are given an ADHD diagnosis – to focus on the task in hand. Indeed, despite claims to the contrary, for the many dozens of substances marketed as smart drugs, from natural products like ginseng and gingko to hormones such as DHEA and even hazardous pharmaceuticals such as hydrazine, it is doubtful that such a purely cognitive effect is possible, such is the complexity of neural processes and brain regions engaged in each and every cognitive act – from the simplest, like deciding what to have for breakfast to the most complex, like code-breaking.

Unsurprisingly, the use of performance- and attention-enhancing drugs has led to some interesting anomalies and debates amongst ethicists. Why should drugs that the military encourages its pilots to use be banned for athletes by the World Drug Anti-Doping Agency and in the classroom unless the student is given an ADHD diagnosis? And how about the parent who recognizes the edge that Ritalin would give and so presses for an ADHD diagnosis to get the drug prescribed for their child? Tricky, too, for the pilot who is also a brilliant athlete or taking an online university course. As moral philosopher Michael Sandel asks in his book *The Case Against Perfection*, is there a distinction between using such a chemical enhancer and taking extra coaching lessons? When asked what they thought about the use of Ritalin, a group of teenagers responded that they saw using the drug as cheating, and overwhelmingly rejected it, but, when asked the concrete question of what they would do if others in the class were using Ritalin, they reversed their position and said they would too.

An increasingly popular alternative to using drugs, at least in the US, is to charge up the brain electrically. One method requires little more than a pair of electrodes placed over the scalp and two nine-volt batteries to deliver a current of 1–2 mA through the brain. This is transcranial direct current stimulation (tDCS). tDCS is one of a family of electrical and magnetic devices originally developed as experimental approaches to the treatment of psychological and neurological problems, from depression to Parkinson's disease. They quickly became of interest to both the military (to improve and speed up intelligence analysis) and the computer gaming industry, as potential methods of enhancing learning and memory. There is some research suggesting that students given a learning task while being buzzed with tDCS remember that particular task better when tested a few weeks later, though the effect is weak. Although in the US the FDA has refused a licence for the medical uses of tDCS, the kits can be sold as cognitive enhancers through direct-to-consumer marketing, provided they carry a health warning. Depending on the numbers of attached bells and whistles, the web-advertised kits market at anything from $150 to $400.

Neuroeducation and neuromyths

On the desirability of neuroeducation the OECD, the Royal Society and the Wellcome Trust all agree. All, however, are concerned about the spread of what they call neuromyths – that is, widely believed ideas about the brain based on bad, outdated or even non-existent science. Some of these are ideas about how the brain works; others about educational practices and devices claimed to enhance learning. According to the Wellcome survey, these myths are believed by many teachers on their panel, who therefore become soft targets for aggressive marketing. Being interested in science is not the same as knowing about it, as demonstrated by a number of 1992 studies, supported by the Economic and Social Research Council, which found that the more people knew about science the more sceptical they were of its claims.[11] More recent studies indicate that this is not the case for neuroscience. In these studies, most respondents, regardless of their level of knowledge, respond positively to propositions claiming to be based on neuroscience. But without the appeal to neuroscience, these same respondents are less than willing to support the same proposition. Neuroscience seems to confer a greater authority than other life sciences – has it become the location for what was once spoken of confidently as the soul?

When such myths are seen as presenting a significant challenge to orthodox science in any field, then national academies such as the Royal Society see it as one of their responsibilities to police the frontiers, drawing on their authority to specify just what is and what is not science.[12] The origins of some of these popular ideas of how the brain works are often obscure. A good example is the claim that 'we only use 10 per cent of our brain'. Neuroscientists are baffled as to where this commonly held belief came from, or even exactly what it means – that 90 per cent of the brain's 100 billion neurons are inactive or redundant? One of the earliest references seems to be in Dale Carnegie's best-selling *How to Win Friends and Influence People*, in 1936, but where he got the idea from is unknown. Certainly it is not borne out by any neuroscientific evidence which, on the contrary, points towards the almost continuous activity of neurons across the entire brain. The 10 per cent figure may, however, hint at both the idea that there is plenty of reserve brain capacity that people could tap into, if properly trained, and at the brain's plasticity, the ability to remodel synaptic connections and neural pathways.

As with the idea of listening to Mozart while in the womb, the roots of other such ideas are easier to trace. Popular among teachers, particularly those interested in neuroscience, is a procedure marketed by a company called Brain-Gym. This involves breaking a lesson for a period in which children are instructed to stand up, place thumb and forefinger on the soft spot under the collarbone and rub gently. The process is said to increase blood flow to the brain and therefore enhance the learning potential of the class. Despite its improbability, and lack of any physiological evidence, this bizarre activity probably finds its origin in the neuroscientific and cognitive psychological observations that: (a) the brain is hugely greedy for oxygen; (b) exercise increases blood flow; and (c) breaking a learning period for brief exercise may help concentration when studying is resumed.

Another pair of myths, deeply rooted in everyday speech and given credence by many teachers in the Wellcome survey, link the left-brain/right-brain distinction with beliefs about inborn differences between men and women. The claim that men's brains differ from, and are larger than, women's has venerable antecedents; in Western thought it goes back to Aristotle. It was a taken-for-granted assertion in the scientific literature of the nineteenth century (it appears, for instance, in Darwin's writing) and was vigorously deployed in the attempts to exclude women from the universities. While most critical discussion of IQ theory in the post-1945 UK context focused on the sheer improbability of constructing a culture-free test, and thus its systematic bias against working-class children, the gendered statistical manipulation of IQ test results has received less attention. In the 1970s, a re-examination of past test scores ordered by Frances Morrell, Chair of the Inner London Education Authority, discovered that the girls had scored higher than the boys; however, as the patriarchal educational psychologists in charge of the testing considered this to be impossible, they adjusted the scores and the questions to produce equality.[13]

Contemporary neuroanatomy disputes the claim that there are gross average differences in size between men's and women's brains when corrected for differences in body size, but accepts that there are differences in some internal brain structures and biochemistry. There are – disputed – claims that there are differences between men and women in the thickness of the corpus callosum, the great tract of nervous connections between the hemispheres, providing a brain-based suggestion as to why men are – allegedly – more single-minded and women more able to multi-task. Other differences lie in the response of neurons deep in the brain to hormones such as testosterone. The autism expert Simon Baron-Cohen has argued that it is the interaction of testosterone with receptor molecules in a particular region of the brain during foetal life that masculinizes an otherwise female brain, and it is this that determines an 'essential difference' between the sexes, including making males more susceptible to autism.[14] He terms this 'brain-organization theory', but both his methods and overdrawn inferences concerning cognitive and emotional differences between men and women have been strongly criticized by psychologist Cordelia Fine[15] and sociomedical scientist Rebecca Jordan Young.[16]

As for hemispheric differences, in popular discourse the left brain is supposed to be cognitive, linear, masculine; the right brain affective, visual, feminine. The distinction has fostered the belief – denounced by the Royal Society and others but taught to and popular amongst teachers, with an abundance of web and printed material endorsing it – that children are naturally left- or right-brained, and that such brain differences will determine an individual child's learning style, described as primarily visual, auditory or kinaesthetic (VAK). Teaching strategies should therefore be matched to each pupil's appropriate learning style. Visual learning is said to be right brain, auditory to be left brain, and kinaesthetic is presumably when both are equally engaged. According to at least one VAK website, learning styles are genetically determined and can be deduced from where a person looks when thinking (visual learners look up, auditory learners straight ahead, and kinaesthetic learners look down). The claim that teaching children according to their supposedly preferred learning style improves performance is not supported by evidence. Yet a comparative study of UK and Dutch teachers interested in neuroscience, and therefore who might be expected to have more accurate knowledge, found that 93 per cent of UK and 96 per cent of Dutch teachers believed in the importance of VAK learning styles; 91 and 96 per cent believed in the relevance of left/right brain differences to teaching and learning.[17]

But the origins of some of these 'myths' can often be found within neuroscience's not-so-distant past. The idea that the brain's part in cognition, emotion and indeed learning styles is partitioned between the left and right hemispheres probably dates from the split-brain experiments of Roger Sperry in the 1950s. Sperry studied patients suffering from intractable epilepsy, who had been treated by severing the corpus callosum, a procedure aimed at preventing the spread of the electrical storm of epilepsy from one to the other hemisphere – and which succeeded. He and his colleagues found that when the hemispheres could not communicate through the callosum, each responded independently and differently to features in the patient's environment: the left to spoken words; the right to visual cues. But for people with normally functioning brains, and thus with left and right hemispheres in continuous and coordinated communication, the partitioning of function between left and right brain is irrelevant to performance.

In this, as in other cases, today's neuromyth is based on yesterday's cutting-edge science.

Neuroeducation within the boundaries

With the myths debunked and the boundaries of good science firmly delineated as in the Royal Society's report, the neuroscientists can then make the case for the contribution of their discipline to education. A good example is *The Learning Brain: Lessons for*

The OECD report's executive summary lists eight 'key messages and themes for the future'. They are:

1. Educational neuroscience is generating valuable new knowledge to inform educational policy and practice.
2. Brain research provides important neuroscientific evidence to support the broad aim of lifelong learning.
3. Neuroscience buttresses support for education's wider benefits, especially for an ageing population.
4. The need for holistic approaches based on the interdependence of body and mind, the emotional and the cognitive.
5. Understanding adolescence – high horsepower, poor steering.
6. Better informing the curriculum and education's phases and levels with neuroscientific insights.
7. Ensuring neuroscience's contribution to major learning challenges (dyslexia, dyscalculia and dementia).
8. More personalized assessment to improve learning.

Education, by the neuroscientists Sarah-Jayne Blakemore and Uta Frith. As they write:

> It might be hazardous to suggest that educational research itself does not or could not provide the best approach to many educational issues from its own resources and sound (*sic*) scientific thinking. As well as

> asking how neuroscience can inform education, it might be useful to think about how brain science challenges commonsense views about teaching and learning.[18]

However, what they mean by 'commonsense views about learning and teaching' is left obscure; they offer no examples which neuroscience challenges and make few suggestions as to how this might connect to the classroom. Although they speak of the need for a common language between educators and neuroscientists, their glossary contains only neurowords. What is missing from the book is any discussion of the research base underpinning educational theory and practice provided by the humanities and social sciences. A second lack is any programme for engagement with the several publics involved in education: the teachers, parents and children (of an age to understand the issues). Instead, in lockstep with the OECD and Royal Society reports, they address teachers with the unchallengeable authority of neuroscience.

The first three of the OECD's recommendations do no more than plant a flag for neuroscience; the fourth is a sort of comfort blanket for those uneasy with the biologization of education. Only the final set suggests areas in which neuroscience, or neuroscientists, might actually contribute. The Royal Society's recommendations are similar and follow its 'six key insights' into the brain's role in education. Yet there is one that they omit, even though it has immediate relevance to children's capacity to learn. The brain, as every neuroscientist knows, demands more energy than any other body organ and, if their brains are energy-deficient, children will not be able to learn well. The number of children receiving free school meals is steadily rising, projected to reach one in three in the near future, and many children do not have breakfast because their parents rely on increasingly inadequate benefits or through personal problems simply can't cope. Either way the children go to school hungry.[19] One study of pupils taking the English Baccalaureate notes that only 9.7 per cent of those eligible for free school meals are successful compared with 26.6 per cent of the rest.[20] The Royal Society's neuroeducationalists might also have spoken of the need for a free breakfast programme – it's hard to learn on an empty stomach.

Unsurprisingly, not all neuroscientists are swept away by the Royal Society's and OECD's recommendations. Their scepticism about the readiness and relevance of the current state of the neurosciences to the classroom more rarely surfaces in the media – unless as with the Human Brain Project there are huge resources at stake.

A gem among these private dissident observations on the Royal Society Report was that of Vincent Walsh, a prominent neuroscientist studying the effects of magnetic and electrical stimulation on the human brain, shown in his italicized comments below:

1 Both nature and nurture affect the learning brain. *Cross out brain and we lose nothing – nature and nurture affect learning.*
2 The brain is plastic. *Let's try 'People can learn and change.' True, we underestimate the capacity for lifelong change, but this is not a key insight.*
3 The brain's response to reward is influenced by expectations and uncertainty. *Replace 'The brain's' with 'People's' and we have something a teacher can work with (but probably already knows).*
4 The brain has mechanisms for self-regulation. *Let's try 'People can learn behavioural methods of self-regulation.'*
5 Education is a powerful form of cognitive enhancement. *Education IS cognitive enhancement. 'Education is a powerful form of … er … education.'*
6 There are individual differences in learning ability with a basis in the brain. *We can change either end of this one. 'There are individual differences in*

> *learning ability.' Or, 'There are individual differences in learning ability with a basis in economics, class and opportunity.'*[21]

Putting to one side for the present the potential role of neurosciences in understanding and treating neurodevelopmental learning disorders, what practical proposals might one consider for enhancing learning for most schoolchildren? Some insight is provided by looking at the grants awarded under the joint auspices of the Wellcome Trust and the Educational Endowment Foundation in 2014. The EEF is a charity whose aims are to break the links between family income and educational achievement, ensuring that children from all backgrounds can fulfil their potential and make the most of their talents. The EEF's trustees are private-equity company philanthropists, who are also presumably the primary sources of its income. The £6 million initiative funded six projects. Two of these – one on spaced learning, the other on changing the school starting time for teenage pupils – are perhaps indicative of the extent to which, as the OECD recommendations put it, 'educational neuroscience is generating valuable new knowledge to inform educational policy and practice'.

The ethics of research into education

The ethics of educational research are well established within the EEF's programme but are new to the Wellcome Trust, most of whose funded research projects are biomedical, and where informed consent has long been the gold standard. Wellcome has not developed its own guidelines but has adopted those of the EEF, which are less stringent on informed consent than those for Wellcome's biomedical grants. Thus the EEF guide states:

> it is a general principle of research that participants should give informed consent to take part. However, schools routinely innovate, try out new approaches and informally evaluate them all the time. You should use your own judgement and usual process when it comes to deciding whether to gain consent for children to take part in either the intervention or the testing.[22]

However, leaving researchers to decide whether informed consent is appropriate risks a conflict of interest, as we discuss below. More sadly, it is in direct conflict with the 1989 UN Convention on the rights of the child, which 'changed the way children are viewed and treated – in other words, as human beings with a distinct set of rights instead of as passive objects of care and charity'. By contrast, the British Educational Research Association's ethical guidance explicitly acknowledges the UN Convention and defends the right of the child to agency where that child is old enough to make decisions – for example, whether they do or do not wish to take part in a research project.

Spaced learning

As far back as the nineteenth century, psychologists had observed that the most effective way to memorize some material – say, a list of words or the times table – was to study it repeatedly but over an interval, with breaks, even of days, between study sessions. Without drawing on such psychological literature, this is one long-known and successful learning strategy. The spaced learning project, however, draws not on this familiar feature of human learning and teaching, but on a rather speculative extrapolation from laboratory memory studies in species as diverse as fruit flies and mice. The proposed protocol requires breaking up the traditional forty-five-minute class period into three ten-minute sessions, interspersed with distraction breaks during which students juggle, practise basketball or play with modelling clay. The first session introduces the content to be learned or revised, the second repeats it more or less exactly, and the third involves the students in some activity relevant to the content. The current project is derived from a teaching experiment by Paul Kelley, at that time (2007) head teacher of Monkseaton School in the north of England, inspired by an article by the NIH neuroscientist Doug Fields, entitled 'Making Memories Stick', published in *Scientific American*.[23]

Here's the neuroscience on which Fields' article and Kelley's project was based. According to the current neuroscientific hypothesis, memories are stored in the brain in the form of changed patterns of synaptic connections between neurons. Such changes can be observed in the lab when animals are trained on some novel task. Fruit flies have a good sense of smell. If the flies are shocked electrically in the presence of a particular strong scent that they might normally prefer then, given the choice of flying towards a source of that scent or a different one, they will avoid the one associated with the shock. In neuroscience jargon, they have learned the scent–shock association and therefore the avoidance response. It requires several repetitions of the pairing of the scent and shock for the flies to learn the avoidance. The trials can be given in rapid succession (massed training) or in batches separated by rest intervals (spaced), and it turns out that the flies remember better in the spaced than the massed version of the task. A similar difference in memory retention between massed and spaced trials was later found in learning experiments in mice.

So far, so good. The neuroscience is solid enough, and Fields ends his article by speculating that similar learning patterns might occur in humans too. But the jump between a fly learning to avoid a particular scent and a schoolchild studying, say, biology at Monkseaton, is huge. Just for starters, how does remembering a single association between a scent and a shock relate to learning and remembering what hormones are and how they work in a biology lesson? Does it matter that in the animal experiments the flies are learning to avoid an unpleasant experience, whereas the teacher is working with the students, rather than shocking them, to help them acquire meaningful knowledge? Using the same word, learning, to describe both what flies and students are doing may indicate that the underlying brain and cellular mechanisms are identical – or it may not. After all, both humans and computers are said to possess memory, but the coincidence of language is metaphorical; it doesn't imply identity of process. How does the time interval in spaced learning in drosophila scale up in choosing the spacing between repeats of the lesson in the classroom?

Nonetheless, head teacher Kelley and his colleagues pressed ahead, conducting three sets of spaced learning experiments[24] with separate cohorts of thirteen- to fifteen-year-old students studying the GCSE biology syllabus,[25] having 'obtained the informed consent of both pupils and their parents'. (That the researchers do not report the percentage of those consenting would concern most social scientists as to whether pupils or their parents felt they were in a situation where refusing, given the institutional power of the teacher/researcher, was unwise.) Some days later, the students were tested for how well they remembered the material in a 'high-stakes' multiple-choice test, and their results were compared with students who had been taught the same material conventionally over a four-month period. Both normally taught and spaced-taught students scored better than if they had only given random answers on the multiple-choice test – scarcely surprisingly – but there was no difference in scores between the two taught groups. Nonetheless, the researchers claimed success, concluding that the students could learn as much in one hour of spaced learning as in four months of conventional teaching.

As one student described it: 'the review of my whole biology unit was completed in about twelve minutes. The nervous system, diet deficiencies, hormones and the menstrual cycle, drugs, and defence from pathogens all whizz by on slides shown at the dizzying rate of seven to eight per minute ... In the end I am left with a movie in my head of the lesson.'[26] Or, perhaps, as Woody Allen put it: 'I took a speedreading course and read *War and Peace* in twenty minutes. It involves Russia.' As much educational and behavioural research has shown, learning by way of repetition with intervals can be one of several effective strategies, as the recent book *Make it Stick* by cognitive psychologist Peter Brown and colleagues points out,[27] but investing it with a rhetorical neuroscientific base is yet again a bridge too far.

Teen sleep

A second project funded under the Wellcome/EEF aegis concerns the teen brain, whose immaturity has since the 1990s been seen by neuroscience as the biological base for the erratic, rebellious and risky behaviour of teenagers. One of these is the well-recognized phenomenon of teenagers staying up late, finding it hard to get up in the morning and having difficulty in concentrating at school. Given the PISA rankings, this has become a political problem demanding research to develop policies that, by accommodating the teen brain, helps teenagers modify their behaviour. The research is intended to assess the effect of later school starting times, coupled with a sleep education programme, on teenagers' educational achievement.

The neuroscience behind the teen-brain concept is built around the findings that, although the great proportion of brain development has taken place by puberty, some regions of the brain – notably the prefrontal cortex, a great mass of tissue located just behind the forehead – do not reach their full growth until a person is in their twenties. The PFC is a region associated with planning complex cognitive behaviours, decision-making and moderating social behaviour. Hence the claim that an adolescent's immature brain is one of – as the OECD somewhat pejoratively puts it – high horsepower but poor steering. Teenagers may look physically mature but their undeveloped PFC explains why they might 'take risks, be impulsive, emotional, rebellious, disorganized, distracted and late', according to the Solihull Approach discussed in Chapter 3. The OECD's metaphor is routinely echoed, though more sedately, in many studies of the teenage brain.

Sleep researchers suggest that, perhaps as part of coming to terms with the physiological and emotional transition from childhood, adolescents need around nine hours' sleep a night. Other neuroscientists offer an alternative explanation. In adolescence the circadian rhythm, which describes a person's pattern of sleep/wakefulness, is shifted towards 'eveningness'; that is, going to bed and waking up late is an intrinsic feature of adolescent neurobiology – hence 'teen sleep' – albeit not one characterizing the working-class young of nineteenth- and earlier twentieth-century industrial Britain, with the need to clock in at the factory gate.

Historians and sociologists have a different narrative: they note that the category of the 'teenager' did not exist at the beginning of the twentieth century. There were adolescents and they were in their teens, but the new identity of the teenager with increasing cultural independence from adult mores came into existence in the 1920s, was in general use in the US by the 1940s and, within a decade, in the UK. Part of the explanation for the new identity lies in the increased wealth of industrialized countries. With the ability of parents to give their adolescents increasingly generous pocket money, the market soon saw that the teen dollar offered an opportunity. A teenage culture was born, two of its major icons being the short-lived James Dean of *Rebel*

Without a Cause, which captured the teenagers' angst, and Elvis Presley, who broke the barriers between black and white music, while also introducing sexualized dancing plus a distinctive new dress code. As the sales of his records mounted, the clothes industry assiduously and successfully marketed the Presley style. With this, the biological adolescent also became the cultural teenager. That said, changes in behaviour, as well as changes in cultural forms and values, are usually matched by changes in the brain.

In consequence of their commitment to neuro-izing teenage sleep patterns, the Wellcome project is to study the effect of altering the school start time for adolescents from the standard UK school hour of 9 to 10 a.m. (Interestingly, Paul Kelley, the head teacher who introduced spaced learning at Monkseaton School, also trialled postponing the school start time from 9 to 10,[28] and is currently attached to the Wellcome-funded research teams experimenting with both spaced learning and teen sleep.) Even as this research was being proposed in England, however, the practice had been widely adopted in the US. By 2010, forty-three schools in forty-three US states had delayed their start times – but with one significant difference from the UK proposal. In the US, school starts are often as early as 7.15, and the delayed starts vary from 7.45 to 8.30. In 2014, the American Academy of Pediatrics formally recommended a start time of 8.30 – that is, the adolescent circadian delay in the US is supposed to end half an hour before the normal UK start time, and an hour and a half before the proposed delayed UK start time!

It is hard to imagine any neuroscientific rationale for this transatlantic difference, except insofar as US culture changes teenage brains differently from European. But this attempt by neuroscientists to accommodate teenage 'eveningness', however caused, should be viewed sympathetically as a first attempt to adapt the social world to what neuroscientists understand as the teenagers' biological needs. Hard going in this 24/7 culture – and, practically, given that some might for preference sleep till noon, who is to wake the teenager up even at the later time when Mum and Dad have gone to work?

These two projects – spaced learning and teen sleep – are intended to advance evidence-based educational practice and are designed on the quantitative model of random control trials (RCT), followed by meta-analysis of all similar studies, pioneered by doctor and health researcher Archie Cochrane. Their success will presumably be measured in terms of exam results for the participants. However, this is but one measure; qualitative research also has a place, and RCTs are not the only tool by which to evaluate successful practice. As public-health specialist Mark Petticrew and medical sociologist Helen Roberts argue,[29] there is no hierarchy of evidence: methods may include quantitative, like RCTs, cost–benefit analysis, and qualitative studies, or some appropriate combination. The choice of a research method for evaluation of intervention must depend on the question being asked. Devoid of qualitative research, such as Choudhury's study of teens to which we turn below, spaced learning and teen sleep may tick only quantitative boxes.

And the view of the teens?

Teenagers, asleep or awake, have become one focus of the deluge of neurotalk in popular science books, radio and television programmes, which confidently locates teenagers' risky behaviour, from drug use to road incidents, teen pregnancy and sexually transmitted diseases, as in their brains. The audience this popularized neurotalk seeks to address concerns not only parents and teachers, as in the past, but now the teenagers themselves. The assumption is that learning about how their own brain works, with its slow-developing pre-frontal cortex and its plasticity, will provide teenagers with a new way of thinking about themselves – of constructing new subjectivities. Critical neuroscientist Suparna Choudhury and her colleagues[30] explored how teenage girls in a London school drew on the discourse of neuroscience – if at all. Far from being uncritical sponges soaking up the neurotalk, 76 per cent of the girls perceived the 'teen self' of the neuroscientists and amplified by the media as stereotypical and failing to see diversity. They considered the concept failed to explain why so many work hard most of the time, are fast developing their social skills, are reasonably well organized and achieve many of their goals. They insisted that it was possible to both party and get straight As – it wasn't an either/or choice. Stereotyping was, the researchers observed, 'a trope that wended its way through the discussions' and, so far as the teenagers were concerned, the stereotypes employed about them were overwhelmingly negative.

The teens explained behaviour and mental health issues in terms of personal and social experience with almost no reference to biology – except for hormones. If the team writing the OECD education report, with its categorization of teenagers as having 'high horsepower, low steerage', had engaged in discussion with such teens their report would surely have been less patronizing and one-dimensional. It seems that the OECD and many well-meaning neuroscientists still live in a one-way street.

Neuroscience and neurodiversity

The source of the learning difficulties that many children face is located in the world into which they are born. Poverty and insecure housing, parents with problems of their own to contend with, and hunger, all militate against the tranquillity, interest and aspiration which education demands. But, even against this general background, it is increasingly recognized that a small subset of children have specific learning difficulties in literacy, numerosity and in the capacity to understand and relate to others – difficulties formalized into categories, as dyslexia, dyscalculia and autism. Devised by educationalists and neuroscientists to categorize people with 'atypical' cognition, the labels are beginning to be embraced by the subjects themselves under the banner of neurodiversity, by contrast with the neurotypical majority. Neurodiversity respects the fact that not everyone's brain works the same way, and it's not hard to find people describing themselves or one another as 'a bit Aspie' or 'a bit dyslexic'.

The numbers of the neurodiverse are hard to estimate, as categories and boundaries are constantly shifting and the diagnostic criteria for describing a child as dyslexic, dyscalculic or autistic are based largely on educational and behavioural observation. The British Dyslexia Association website suggests up to 10 per cent of children and adults may have reading difficulties associated with dyslexia. Dyscalculia is estimated as affecting 6 per cent. According to the UK National Autistic Society, autism spectrum disorder (a category that embraces what was previously called Asperger's syndrome) affects just over 1 per cent. There are no clear biomarkers, although associations have been found with some rare gene variants. Although the reductionist drive of neuroscience and genetics powers the search for simple biological explanations, the complexity and variety of the ways of thinking and behaving covered by these unifying labels can make the diagnoses contentious. What, then, can educational neuroscience offer, diagnostically or in terms of support, for the children themselves, their parents and teachers?

One feature is that most of these diagnoses are commoner in boys than girls – dyslexia, for example, being up to five times more frequent. Behaviour geneticists, ignoring the effects of social gendering, then attribute these differences to the genetic difference between the sexes, females having two X chromosomes, males (XY) only one. If one of the X chromosomes in females is normal and the other carries the gene variants that may predispose towards the condition, the normal may override the variant. In the absence of the second, normal X, this option is not available to males – part of what Baron-Cohen calls 'the essential difference'.

Dyslexia

Because none of the gene variants is strongly predictive of these learning difficulties, the current emphasis amongst cognitive neuroscientists is to see autism, dyslexia and dyscalculia as neurodevelopmental disorders – that is, they result from a combination of many gene variants, epigenetic and environmental factors during the formative years of child development, rather than having genetic lines of causation. An additional problem for genetic theories of causation lies in the epidemiological evidence that dyslexia is more often diagnosed in middle-class than working-class children, a class difference that has made some suspicious that it is a polite way of accounting for a middle-class child's poor performance.

Neuroscientists set these concerns apart. For them, the focus is on identifying the brain mechanisms underlying the reading and writing problems. It's been known since the nineteenth century that damage to particular left-hemisphere brain regions can affect comprehension and speech, and neurologists went on to identify a wide variety of specific problems; alexia (reading difficulties), agraphia (writing difficulties) and apraxia (difficulties with speech). It wasn't, however, until the advent of fMRI that it became possible to identify regions of the brain activated during the various sub-routines engaged during reading. In adults, normal reading engages a network of neurons in the frontal and temporal regions of the left hemisphere, called the visual word-form area. Identifying these brain regions makes it possible for researchers to ask – adopting a classical biomedical strategy – how they respond in children who have reading problems. Without going in detail into the complexities of the research, the broad and perhaps unsurprising conclusion is that, compared with usual developing readers, children with dyslexia show reduced or different activity in some of these regions.[31] Older people with dyslexia develop compensatory strategies involving different brain regions.

The next step for the imagers was to see whether and how teaching might both improve literacy and modify brain responses. In one interventive study, dyslexic children aged seven and upwards were given intensive daily remedial teaching over a period of eight months, focusing on phonology and orthography. Their reading improved, and the level of neural activity in the visual word-form area also increased, showing that the functional connectivity of the brain can be reshaped, to approach that found in more fluent readers.[32] The researchers argue that without the remediation, the compensatory strategies that dyslexic people develop require more and wider regions of brain activity, and hence more effortful. This correlation casts further light on the brain mechanisms involved in reading, which may be malfunctioning in dyslexia, but, as the researchers themselves acknowledge, provides no guidance as to how these neuroscientific findings might improve the children's reading ability, over and above the intensive remedial teaching.

Dyscalculia

Some appreciation of numbers and the ability to count are important in most cultures, despite wide variation in the methods people use, as studied by anthropologists. Nor is this limited to humans. Several non-human species, including not merely our near-evolutionary neighbours the primates, but also pigeons and chickens, can also count – as shown, for example, by teaching pigeons to peck five times at a coloured light for a food reward. Numbering serves a variety of purposes. The most obvious is termed 'numerosity', that is, to be able to count the numbers of things in a set, for example five fingers on a hand, and their relative quantity – five apples are more than four. Another purpose is to define order, as in the page numbers of a book – page 100 isn't greater than page 99, but is placed after it in sequence.

As with literacy, psychologists and educators alike have long studied how children learn maths, and how the capacities for numerosity and ordering develop can be followed even in babies by observing how long they look at objects or displays. Babies respond to novelty by looking for longer at a novel object or pattern than they would at familiar ones. If week-old babies are first shown two identical objects, and then more are added, the babies register the difference in number by paying attention to the novelty. That is, they can distinguish numerosities. How about the brain regions involved? As summarized by cognitive neuroscientist

Brian Butterworth,[33] in adults, three brain areas in the frontal and parietal lobes are particularly important for numbers and arithmetic, and damage to them can impair calculation. The brain regions involved in distinguishing numerosity in young babies are studied by recording brain responses using non-invasive methods such as the EEG hairnets described in Chapter 3. When three-month-old babies are first shown a display of two objects and then a third is added, there is a change of signal in the right parietal lobe, the same region that is used in processing numerosities in adults.

Mathematical performance varies widely amongst children and adults in and between developed countries, as the PISA rankings show. So too do teaching methods and resource investment as, say, between China and Britain. Describing someone with low mathematical achievement as having a specific disability, dyscalculia, is a relatively recent addition to the catalogue of learning difficulties. Butterworth sees dyscalculia as a neurodevelopmental disorder, partly inherited and caused by a core problem in processing numerosities, resulting in low arithmetical achievement. When their brains are imaged during mathematical tests, eleven-year-old dyscalculic children show lower activation of the frontal and parietal lobes than do their averagely performing peers. As with the literacy studies, there are suggestions that imaging might have predictive value and at some time in the future may point to developing specific remedial strategies.

Fascinating to neuroscientists as such findings are, and enthusiastic as many teachers are about the relevance of their observations to classroom teaching, it is important to view these enthusiasms cautiously. Even if we accept that dyslexia and dyscalculia have specific neurological correlates, does knowing those correlates help in designing a teaching strategy? Take the remedial programme in dyslexia described above, involving intensive one-on-one tuition to learn to read. It required no neurological understanding to develop the programme, and its success does not depend on whether the dyslexic child's developmental problem is located in the brain or the big toe. As Bruer argued two decades ago, it is possible to build bridges between neuroscience and cognitive psychology, and between cognitive psychology and education, but between neuroscience and education is still a bridge too far. The most powerful argument for the role of neuroscience may well be that the finding of specific brain differences makes it possible to see these conditions as 'real'. In such situations, the claim that neuroeducation is a help to educators needs standing on its head. When teachers identify children with specific learning difficulties, they inadvertently provide neuroscientists with an entry into understanding the brain mechanisms involved in literacy or numeracy. The signposts on the one-way street that neuroeducation offers, of educating teachers about the brain, need turning round, enabling teachers not merely to collaborate with but to educate neuroscientists.

Public engagement with neuroscience

Just because this book is critical, it would be a serious mistake to fail to recognize that this is pretty much a golden age for neuroscientists – at least those with jobs and grants. Young neuroscientists, despite their skill and long training, are not immune from being pushed into the precariat. Technologies which make the workings of the living brain visible and manipulable, from single genes, synapses and neurons to the entire neural system, are primed both to answer old questions and suggest entirely new and previously inconceivable ones. With the new neurotechnologies, bolstered by the Euro-American mega-projects of HBP and BRAIN, also comes the promise of power, the imaginaries of predicting behaviour, mending broken brains, transforming working ones, reading thoughts and intentions, and even creating computer brains for the new generation of robots. Small wonder, then, that many want to share the intense pleasure they get from their work with those outside this arcane but exciting world.

Today, in good measure because of the crisis in trust in science occasioned by recurrent disasters and scandals, from mad cow disease to MMR, the British scientific establishment, from the Royal Society to the professional organizations, encourages scientists (well, not those working in defence) to write books, present science programmes and support science festivals. Scientists as experts, as sociologist Harry Collins insists,[1] remain respected, but misplaced deference has retreated, replaced by a lively multiplicity of publics very willing to engage with the knowledge on offer, above all when it is knowledge which bears on their lived experience. But, since its heyday in the 'oughties', public engagement has become thin gruel. The concept of 'engagement' has been diluted beyond recognition and often appropriated by what should be more accurately called 'public relations'. Universities put on free lectures, science open days and exhibitions of art/sci. Opening their doors to the public that pays for them is only to be welcomed, but to put these activities – as they often are – under their programme of 'public engagement' is to fly a false flag. The public are invited in as guests, audience, or viewers, but what they are not being offered is anything like the *Meeting of Minds*, which, as we described in Chapter 2, claimed to have provided an 'unprecedented opportunity to give ordinary people a role in guiding … policy development in a complex scientific field … a breakthrough in participatory government'.

As neuroscience expands into early intervention and education, it needs to show some humility for the expertise of others in collaborating with the existing fields and disciplines integral to education studies and the very considerable body of research, chiefly in the humanities and the social sciences, that underpins them. Further, neuroscientists must engage with the range of publics germane to specific projects. Thus, early intervention projects would require engaging with the parents, carers, nursery staff and, where the children are able, them too. Engagement would involve listening to the experiences and insights of these as participants in the shaping of the research design, not as passive subjects or recipients of information about it, but with their right to withdraw or take part protected.

Hope, hype and neoliberalism

Like other life sciences that have preceded it, notably genetics, advances in understanding the brain have been accompanied by hope and hype, amplified by a compliant media. In today's neoliberal economy, both comply with, and help create, the demands of neoliberal societies. Neuroscience's methodological focus on the individual brain is in accord with that of neoliberalism on the individual rather than the collective, and with its public-policy initiatives emphasizing self-reliance, aspiration and the will to succeed. In this economy, brains (not the children that surround them) as the repository of mental capital are seen as a resource, and parents are required to lift their children out of poverty by way of their neurons and the magic of brain plasticity. Often poorly understood or over-extrapolated neuroscientific insights are co-opted in support of public policies for early intervention projects, including the packages offered by the many private-sector players. Schools are besieged by unsolicited advertising for the dubious merits of brain gyms, neurotraining programmes and VAK learning styles.

Helping neurodiverse children thrive in a scientific and technological world demanding literacy and numeracy is important, and neuroscientists studying children with diverse brains are making an increasing contribution, but the prerequisite is the recognition and acceptance of neurodiversity. The neuroscience-based proposal to modify the social context to accommodate what they term the 'eveningness' of the teen brain, by changing school hours to a later starting time, has to be seen as a pro-teenager move, though the research outcomes are still awaited. Although not entirely unproblematic, given the problems of labelling and medicalization, these are promising examples of the positive use of neuroscience to tackle social and political problems and issues. Such examples underscore the need for more neuroscientists to position themselves on the side of the neurodiverse they study, willing to consider (and hopefully help modify) the social world in ways that could empower their subjects.

Pessimism of the intellect and optimism of the will?

But these micro-interventions pale into insignificance when faced with the huge problems of poverty and inequality. It's not necessary to understand the workings of the brain to know that precariously housed, underfed children find it hard to study and learn. And despite the government having redefined poverty by ignoring income, instead invoking worklessness and general fecklessness, the 1 per cent get steadily richer, while the poor are driven deeper into poverty, stigmatized for circumstances over which they have little or no control. In Britain, cuts in benefits and proposed but delayed abolition of tax credits are accompanied by the growth of food banks and the numbers of children who go to school breakfastless. Within this political economy and the huge inequalities it generates, a tiny minority of children are increasingly privileged, growing up to think of themselves as entitled by right, while the children of the precariat are made poorer and less confident in themselves. Unrestrained capitalism is apparently not to blame; instead, its ideology blames the parents – they are deficient, lack mental capital, have weak parenting skills, are insufficiently ambitious for their children and fail to see the importance of education for them. To stop these children becoming a burden on the state, the Tories loudly demand that Something Must Be Done. From the neurosciences, insights, real and imagined, are invoked both to explain these moral deficits and to devise programmes to compensate for them. Interventive programmes with long and deep funding may well help a few, irrespective of the often mythic neuroscientific assumptions which underlie them, but the majority will continue to miss out. The language of public policy speaks of targeting the excluded – not the language of universalism and solidarity.

We began this book with a question – one that can be read in several different ways: can neuroscience change our minds? In charting the rise of this new technoscience and its public-policy ambitions, our implicit answer has been necessarily ambivalent. To say both yes and no is not a cop-out but a restatement of our overall perspective of the mutual shaping of science and society. In the most literal reading of our question, a science which produces chemical and physical means to alter thought and feeling manifestly changes our minds. More subtly, like other technosciences, neuroscience is both affected by and modifies our culture, and hence our consciousness. In the meantime, as our critique of neuroscience's reach into child development and education has insisted, first, despite the contribution of neuroscience to understanding brain development and diversity, the problems are primarily social and economic. Second, however well-meaning its advocates, neuroscience alone is unable to mitigate the inequality and deprivation that are an integral part of an intensely marketized economy. Social and political collective understanding and action – despite the challenges – offer the only way forward.

NOTES

Introduction

[1] O'Connor, C. O., Rees, G. and Joffe, H., Neuroscience in the public sphere, *Neuron* 74: 220–6, 2012.

[2] Jasanoff, S., *The Co-production of Science and the Social Order*, Routledge, 2006.

[3] Discussed in our earlier book, *Genes, Cells and Brains: The Promethean Promises of the New Biology*, Verso, 2012.

[4] Social theory has a very different theory of consciousness, seeing it as a product of social relations.

[5] Healy, D., Conflicting interests in Toronto: anatomy of a controversy at the interface of academia and industry, *Perspectives in Biology & Medicine* 45: 250–63, 2002.

[6] As the term 'neurotechnoscience' is cumbersome, we follow the common shorthand practice of using 'neuroscience' much as we use the singular for the plural 'neurosciences'.

[7] Rapp, R., A child surrounds this brain: the future of neurological difference according to scientists, parents and diagnosed young adults, *Advances in Medical Sociology* 13: 3–26, 2011.

[8] Vidal, F., Brainhood: anthropological figure of modernity, *History of the Human Sciences* 22: 5–36, 2009.

[9] We discussed the rise of evolutionary psychology in our book *Alas, Poor Darwin: Arguments against Evolutionary Psychology*, Cape, 2000.

Chapter 1 The Rise and Rise of the Neurosciences

[1] Gould, S. J., *The Mismeasure of Man*, Norton, 1996.

[2] Crick, F., *The Astonishing Hypothesis: The Scientific Search for the Soul*, Simon and Schuster, 1994.

[3] Greenberg, G., *The Book of Woe: The DSM and the Unmaking of Psychiatry*, Blue Rider Press, 2013.

[4] Sorge, R. E. and twenty others, Olfactory exposure to males, including men, causes stress and related analgesia in rodents, *Nature Methods* 11: 629–32, 2014.

[5] Hyman, S. E., Revolution stalled, *Science Translational Medicine* 4: 155cm11, 2012.

[6] Dendrites and axons are the fibrous projections from the neurons that respectively receive input signals and transmit output signals.

Chapter 2 The Neurosciences Go Mega

[1] Rose, H. and Rose, S., *Genes, Cells and Brains: The Promethean Promises of the New Biology*, Verso, 2012.

[2] Human Brain Project website: overview, consulted October 2015.

[3] Irwin, A. and Wynne, B. (eds), *Misunderstanding Science: The Public Reconstruction of Science and Technology*, Cambridge University Press, 2004.

[4] European Union web archive *European Citizens' Deliberation on Brain Sciences*, consulted July 2014.

[5] Markram, H., Seven challenges for neuroscience, *Functional Neurology* 28: 145–51, 2013.

[6] Human Brain Project website, consulted January 2014.

[7] Calabrese, E. et al., A diffusion MRI tractography connectome of the mouse brain and comparison with neuronal tracer data, *Cerebral Cortex*, doi:10.1093/cerecor/bhv121, 2015.

[8] Sample, I., Scientists digitise rat's brain – and a supercomputer's whiskers twitch, *Guardian*, 9 October 2015.
[9] For a review and introduction to the ethical debate, see Poldrack, R. A. and Farah, M., Progress and challenges in probing the human brain, *Nature* 526: 371–9, 2015.
[10] Hugh Herr (MIT), quoted in *Financial Times*, 23 February 2013.
[11] Waldrop, M. M., Computer modelling: brain in a box, *Nature* 452: 456–8, 2012.
[12] Editorial, *Nature* 511: 125, 2014.
[13] Madelin, R., *No Single Roadmap for Understanding the Human Brain*, European Commission, web archive, 18 July 2014.
[14] Frégnac, Y. and Laurent, G., Where is the brain in the Human Brain Project? *Nature* 513: 27–8, 2014.
[15] Editorial, Rethinking the brain, *Nature* 519: 389, 2015.
[16] Haraway, D., Otherworldly conversations; terran topics; local terms, *Science as Culture* 3: 64–98, 1992.

Chapter 3 Early Intervention: Making the Most of Ourselves in the Twenty-first Century

[1] Foresight, Mental capital and well-being: making the most of ourselves in the 21st century, *Final Project Report*, Government Office for Science, 2008.
[2] Beddington, J. et al., The mental wealth of nations, *Nature* 455: 1057–60, 2008.
[3] For some, notably the psychologist Arthur Jensen, the failure of Headstart to enhance the IQ scores of these children, despite the additional funding, pointed to their genetic cognitive inferiority.
[4] Heckman, J. J., Skill formation and the economics of investing in disadvantaged children, *Science* 312: 1900–2, 2006.
[5] The Conservative government of the time restructured the school boards, excluding women from being elected. Who would want a pioneering interventionist like McMillan around?

[6] Allen, G., *Early Intervention: The Next Steps*, and *Early Intervention: Smart Investment, Massive Savings*, Independent reports, HM Government, 2011.

[7] Cameron, D., *Speech at Demos*, at: <www.demos.co.uk/files/cameronspeech> 2010.

[8] Paton, G., Ofsted: 11,000 childcare places axed in 2010, *Telegraph*, 5 May 2010.

[9] Lloyd, E. and Penn, H., Why do childcare markets fail? Comparing England and the Netherlands, *Public Policy Research* 17/1: 42–8, 2010.

[10] Merrick, B., *Guardian*, 29 September 2015.

[11] Wave Trust and NSPCC, *The 1001 Critical Days: The Importance of Conception to Age Two Period*, The Wave Trust, 2012.

[12] Perry, B. D., Childhood experience and the expression of genetic potential: what childhood neglect tells us about nature and nurture, *Brain and Mind* 3: 79–100, 2002.

[13] Perry, B. D. and Pollard, R., Altered brain development following global neglect in early childhood, *Society for Neuroscience*, Annual meeting abstract, 1997.

[14] Rutter, M. et al., Quasi-autistic patterns following severe early global privation, *Journal of Child Psychology and Psychiatry* 4: 537–49, 1999.

[15] An email from Bruce Perry, 22 April 2014.

[16] Lewis, P. and Boseley, S., Iain Duncan Smith 'distorted' research on childhood neglect and brain size, *Guardian*, 9 April 2010.

[17] Goslet, M., The trouble with Kids' Company, *Spectator*, 14 February 2015; Tryhorn, C., ASA raps 'racist' poster for kids' charity, *Guardian*, 26 August 2009.

[18] *Solihull Approach Resource Pack: The School Years*, 2004, p. 99.

[19] Lumey, L. H., Stein, A. D. and Susser, E., Prenatal famine and adult health, *Annual Review of Public Health* 32, doi: 10.1146/annurev-publhealth-031210-101230, 2011.

[20] Rutter, M., Nature, nurture and development: from evangelism through science towards policy and practice, *Child Development* 73: 1–21, 2002.

[21] Noble, K. G. et al., Family income, parental education and brain structure in children and adolescents, *Nature Neuroscience*, doi:10.1038/nn.3983, 2015.

[22] Gerhardt, S., *Why Love Matters*, Routledge, 2004.

[23] Daly, M. and Wilson, M., *Homicide*, Transaction Press, 1988.

[24] Bruer, J. T., *The Myth of the First Three Years*, The Free Press, 1999.

Chapter 4 Neuroscience Goes to School

[1] OECD, *Understanding the Brain: The Birth of a Learning Science*, 2007.

[2] Goswami, U., Neuroscience and education: from research to practice? *Nature Reviews Neuroscience* 7: 406–13, 2006.

[3] Wellcome Trust, How neuroscience is affecting education, *Report of Teacher and Parent Surveys*, January 2014.

[4] OECD, *The High Cost of Low Educational Performance*, 2010.

[5] Butterworth, B. and Varma, S., Mathematical development, in Mareschal, D., Butterworth, B. and Tolmie, A. (eds), *Educational Neuroscience*, Wiley, 2014, p. 202.

[6] Letters, *Guardian online*, 15 March 2015.

[7] BBC news, Teachers warn of unqualified staff, BBC/news/education-32174423, 4 April 2015.

[8] Press release, Prime Minister's Office, 11 March 2015.

[9] Kintrea, K., St Clair, R. and Houston, M., *The Influence of Parents, Places and Poverty on Educational Attitudes and Aspirations*, Rowntree Foundation, 2011.

[10] Gibbs, S. and Elliott, J., The differential effects of labelling: how do 'dyslexia' and 'reading difficulties' affect teachers' beliefs? *European Journal of Special Needs Education* 30: 323–37, 2015.

[11] Irwin, A. and Wynne, B. (eds), op. cit. (n.3 of Chapter 2).

[12] Howard-Jones, P. A., Neuroscience and education: myths and messages, *Nature Reviews Neuroscience* 15: 817–24, 2014.

[13] Morrell, F., *Children of the Future: The Battle for Britain's Schools*, Hogarth, 1989.

[14] Baron-Cohen, S., *The Essential Difference*, Allen Lane, 2003.

[15] Fine, C., *Delusions of Gender: How Our Minds, Society and Neurosexism Create Difference*, Norton, 2010.

[16] Jordan Young, R. M., *Brainstorm*, Harvard University Press, 2010.

[17] Dekker, S. et al., Neuromyths in education: prevalence of misconceptions and predictors among teachers, *Frontiers in Psychology* 3: 429, 2012.

[18] Blakemore, S.-J. and Frith, U., *The Learning Brain: Lessons for Education*, Blackwell, 2005, p. 7.

[19] The cereal company Kelloggs 2012 report on this problem stimulated widespread public concern.

[20] See: <www.SuttonTrust.com/researcharchive/chain-effects-2015/p7>.

[21] Walsh, V., personal communication, 2014.

[22] Coe, R., Kime, S., Nevill, C. and Coleman, R., *The DIY Evaluation Guide*, Education Endowment Foundation, 2013, p. 9.

[23] Fields, D., Making memories stick, *Scientific American* 292/2: 74–81, 2005.

[24] Kelley, P. and Whatson, T., Making long-term memories in minutes: a spaced learning pattern from memory research in education, *Frontiers in Human Neuroscience* 7: 1–9, 2013.

[25] Disturbingly, while Kelley himself had criticized the MMR/autism link, in the video on their website, of the spaced learning lesson on pathogens and vaccination that Monkseaton School made the teacher is shown linking the MMR vaccine with autism, accessed 20 April 2015.

[26] Kelley and Whatson, op. cit. (n.24 of this chapter, above).

[27] Brown, P. C., Roediger, H. L. and McDaniel, M. A., *Make it Stick: The Science of Successful Learning*, Harvard University Press, 2014.

[28] Kelley, P. et al., Synchronizing education to adolescent biology: 'let teens sleep, start school later', *Learning, Media and Technology*, dx.doi.org/10.1080/17439884.2014.942666, 2015.

[29] Petticrew, M. and Roberts, H., Evidence, hierarchies, and typologies: horses for courses, *Journal of Epidemiology and Community Health* 57/7: 527–9, 2003.

[30] Choudhury, S., McKinney, K. A. and Merten, M., Rebelling against the brain: public engagement with the 'neurological adolescent', *Social Science and Medicine* 24: 565–73, 2012.

[31] Stein, J. and Walsh, V., To see but not to read: the magnocellular theory of dyslexia, *Trends in Neuroscience* 20: 147–52, 1997.

[32] Fern-Pollack, L. and Masterson, J., Literacy development, in *Educational Neuroscience*, op. cit. (n.5 of this chapter, above), quoting Shaywitz et al., 2003.

[33] Butterworth, *Educational Neuroscience*, op. cit. (n.5 of this chapter, above).

Conclusion

[1] Collins, H., *Are We All Scientific Experts Now?* Polity, 2014.

INDEX

Index

Index

Index

Index

Index